STRONG
IN THE RAIN

STRONG
IN THE RAIN

Surviving Japan's Earthquake, Tsunami,
and Fukushima Nuclear Disaster

LUCY BIRMINGHAM
DAVID McNEILL

palgrave
macmillan

First published in 2012 by PALGRAVE MACMILLAN® in the United States—a division of St. Martin's Press LLC, 175 Fifth Avenue, New York, NY 10010.

Where this book is distributed in the UK, Europe and the rest of the world, this is by Palgrave Macmillan, a division of Macmillan Publishers Limited, registered in England, company number 785998, of Houndmills, Basingstoke, Hampshire RG21 6XS.

Palgrave Macmillan is the global academic imprint of the above companies and has companies and representatives throughout the world.

Palgrave® and Macmillan® are registered trademarks in the United States, the United Kingdom, Europe and other countries.

ISBN 978-0-230-34186-9

All maps and charts designed by the Institute for Information Design Japan, which holds all copyright for work on pages viii, 1, 30–31, 50–51, 68, 168, and 194–195.

Library of Congress Cataloging-in-Publication Data

Birmingham, Lucy, 1956–
   Strong in the rain : surviving Japan's earthquake, tsunami, and Fukushima nuclear disaster / Lucy Birmingham and David McNeill.
      p. cm.
   ISBN-13: 978-0-230-34186-9
   ISBN-10: 0-230-34186-1
   1. Tohoku Earthquake and Tsunami, Japan, 2011—Personal narratives. 2. Fukushima Nuclear Disaster, Japan, 2011—Personal narratives. 3. Disaster victims—Japan—Tohoku Region—Biography. 4. Tohoku Region (Japan)—Biography. 5. Tohoku Region (Japan)—Social conditions—21st century. I. McNeill, David. II. Title.
   DS894.385.B57   2012
   952'.110512092—dc23
   [B]
                                                            2012024711

A catalogue record of the book is available from the British Library.

Design by Letra Libre, Inc.

First edition: October 2012

10 9 8 7 6 5 4 3 2 1

Printed in the United States of America.

Dedicated to the people of Tohoku.

To Nanako.

And to Seiya, Nina, Sachi, and Luka,
in the hope they will see a better future.

CONTENTS

Eight pages of photographs appear between pages 102 and 103.

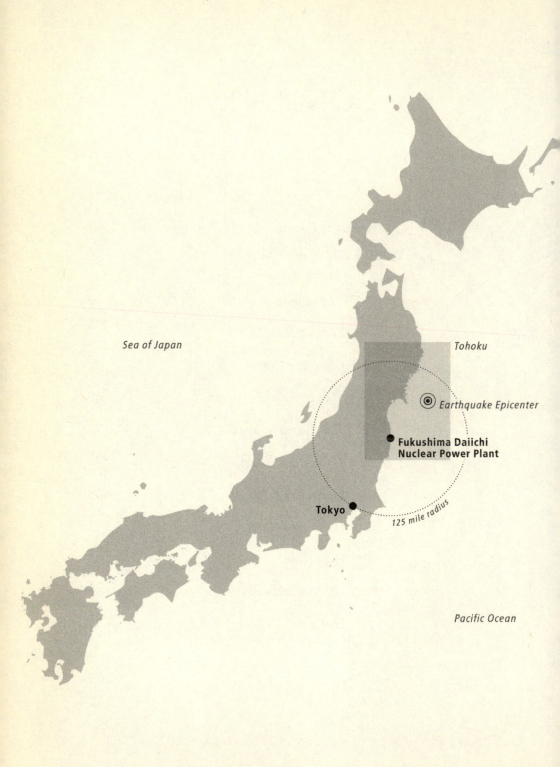

Sea of Japan

Tohoku

⊚ *Earthquake Epicenter*

● **Fukushima Daiichi
Nuclear Power Plant**

Tokyo ●

125 mile radius

Pacific Ocean

Acknowledgments

THE AUTHORS WOULD LIKE TO GRATEFULLY ACKNOWL-
edge the help of the following people:

Mark Selden, Shaun Burnie, Jeff Kingston, Darach
McDonald, and Tom Gill, all of who read and commented
on draft chapters of the book. Liz Maly, Christian Dimmer, Ian Fairlie, Eric Johnston, and Dr. Bruce Parker, who read or commented
on portions of the book. Jake Adelstein, Roland Kelts, and Robert
Whiting, who kindly read the first draft.

Roger Pulvers, who supplied endlessly helpful advice and peer-
less translations of Kenji Miyazawa's poems. Thank you so much
Roger. Nanako Otani, who did many of the translations for the book
and provided useful suggestions. Mamiko Shimizu, Makiko Tazaki
Kimura, Koji Shidara, Koichi Ohtsu, Kei Tanaka, Yasuko Takahashi,
and Michiyo Kato for their skilled translations, guidance, and generous support.

Kate Thomson and Hironori Katagiri for their insights on Tohoku culture and history. Dennis Normile, who offered invaluable
scientific references and opinion. Interviewees Yukari Tachibana and
Chieko Yui for their kind patience. Rachel Vogel, who proposed the
project and helped get it off the ground. Jason Ashlock at Moveable

Type Management, who understood the premise of the book and believed in it enough to make sure it came to life.

At Palgrave Macmillan, our terrific editor Luba Ostashevsky, and Laura Lancaster and Donna Cherry. Hiroshi Sato, who photographed the book's six main characters among many others, introduced us to key contacts, and gave up so much of his time driving us around the Tohoku region. Our remarkable interviewees: Katsunobu Sakurai, Kai Watanabe, Yoshio Ichida, Toru Saito, Setsuko Uwabe and David Chumreonlert, who endured the unendurable, and then agreed to relive it all again so we could record their experiences for this book. Rob Gilhooly, who took some of the most remarkable and moving photographs of the disaster, including the one that adorns the cover of this book. Andreas Schneider, the hugely talented and uncompromising designer of the maps and charts that illustrate the book. Julian Ryall, with whom David McNeill shared an unforgettable three days covering the disaster in Tokyo in March 2011.

To Lucy Birmingham's disaster coverage team at *TIME,* including Howard Chua-Eoan, Zoher Abdoolcarim, Emily Rauhala, and the remarkable Krista Mahr and Hannah Beech, whose bravery and skilled news coverage are a constant inspiration. To David McNeill's editors and colleagues at *The Irish Times, The Independent,* and *The Chronicle of Higher Education,* who provided a forum and testing ground for our work. To our dedicated and hard working colleagues at NHK. David McNeill would like to give a special mention to the brave Daniel Howden from *The Independent,* the first British reporter into the exclusion zone.

Finally, to our long-suffering families and friends who bore the brunt of the inevitably draining and time-consuming work that went into this project. We couldn't have done it without you.

PROLOGUE

March 11, 2011

Someone had better be prepared for the rage.

—*Robert Frost, "Once by the Pacific"*

DAVID MCNEILL, SHINAGAWA, TOKYO

Everyone who lives in Tokyo mentally rehearses where they might be if the Big One strikes. When I first arrived as a student in 1993, I found myself walking through the sprawling, crowded low-ceilinged shopping center of Otemachi and Marunouchi, underneath the city's business district, pondering the apocalypse. The word didn't seem inappropriate: Tokyo has a remarkable, perhaps unique history of almost biblical destruction. In 1923, a 7.9 quake and tsunami famously leveled much of Yokohama and Tokyo, crushing, incinerating, or drowning at least 100,000 people; September 1, the anniversary of that tragedy, is now National Disaster Prevention Day, when millions of schoolchildren practice hiding under desks and evacuating classrooms. Quakes have regularly brought Tokyo and other cities to their knees, and even national icon Mt. Fuji looms threateningly 62 miles away from the capital, ready to spew millions of tons of ash down on the world's largest metropolis.

I've often been asked: How do people with millennia of horrific collective memories manage to repress them and get on with life? One answer is that they don't—at least not completely. The fear of quakes, tsunamis, and volcanoes of course runs very deep among most Japanese. Time and again after March 11, we would hear stories of our interviewees diving for cover, fleeing their houses, or running from the coast the moment the tremors began—instincts that would save lives. I'm thinking now of photographer Kanae Sato, dashing for her car after the quake struck in Ofunato, then driving past neighbors who stayed behind—and died as a result—or fisherman Yoshio Ichida in Soma, leaping from his bath as the shaking began and sprinting for the coast to steer his boat out into the open sea. By the time he returned to port, it had been laid waste.

However, thousands of others ignored ancient precepts about survival in the face of the earthquake and tsunami. In Ofunato, a city with a history of devastating tsunamis, local factory worker Akio Komukai told us about speeding away from the coast after the quake struck and meeting children on their way home from school. "They were walking toward the sea and I rolled down the window of my car and shouted, 'Tsunami tendenko! There's a tsunami coming; you need to run away!'" The young people looked at the 61-year-old Cassandra and said, "Okay, okay." Komukai, who remembers the 1960 tsunami washing away houses, still wonders who among the children survived. "They didn't believe me," he said. "We forget that the sea is close because we build next to it. Then the tsunami comes and washes away the houses and you can see the sea again. And we're reminded."

Each generation builds stone monuments at the highest point of the tsunami that struck their homes, then forgets their lessons, their faded stone lettering a metaphor for collective amnesia.

In the cities, swaddled in technology and the crooning reassurances of government and industry, urbanites can believe that they are safe from natural disaster. Japan has the world's most sophisticated earthquake early-warning system, a combination of high-tech know-how and necessity. With a geology that sits atop four subducting tectonic plates in an area called the Pacific Ring of Fire, the country has over one hundred active and extinct volcanoes and close to 1,500 earthquakes recorded every year.

And urbanites are safe, or safer than most. But two years after I arrived in Japan, there came a reminder that the cost of the collision between primordial seismic instability and life in our most sophisticated cities can still be terribly high. The 1995 Kobe disaster—known as the Great Hanshin Earthquake in Japan—took 6,400 lives, injured 400,000, wiped out 2.5 percent of the nation's GDP, and orphaned dozens of children. Its famous images of toppled highways and Kobe's burning city center brought global humiliation to a country justifiably proud of its postwar construction prowess. I volunteered to help deliver water after that quake and walked around the city, noting how the shaking had left the newer, middle-class suburban housing projects almost untouched but had decimated older, poorer neighborhoods with their wooden frames and heavy tiled roofs. What would happen in Tokyo? I wondered.

On that crisp, sunny afternoon of Friday, March 11, 16 years later, during one of the more pleasant times of the year in the city, I was with my partner, Nanako, in Shinagawa Station, one of Tokyo's biggest train hubs. It was a happy time for us, as we were awaiting the arrival of our baby son, Luka, who was due to arrive in late June. So I found myself in the oldest section of one of Tokyo's busiest train stations with my heavily pregnant partner when the quake struck.

It began not with a jolt, like many quakes, but with an almost lazy undulating rocking motion that slowly built in intensity until the station's roof rattled violently and glass shattered on the platform. A woman somewhere screamed; others clung to husbands, wives, or children. A man ran for the exit and fell over the ticket barrier. I watched as one station attendant sprinted back and forth across the platform, waving his hands in a blind panic and shouting at commuters to stay away from the tracks. It is rare to see Tokyoites panic, and it terrified me. We stood frozen to the spot, hearts thumping violently, and watching the roof, silently praying it wouldn't fall on top of our heads. The tremors seemed to go on forever. Later we would hear that the most intense shaking lasted three to five minutes from what would officially become the Great East Japan Earthquake, hereafter the Tohoku disaster. The Kobe earthquake lasted 20 seconds.

As the shaking subsided and the terror of being buried beneath tons of steel and concrete faded, some people began crying. An almost palpable sense of relief filled the crowded station, like a single, giant sigh. Then everyone began pulling mobile phones out of bags and frantically calling family and friends to check whether they were okay—crashing the network. "Kowakatta!" (That was terrifying!), said one middle-aged woman. "Dame da to omotta" (I thought we were finished), said another. Hundreds of people walked around on wobbly legs, dazed, their schedules shredded, the technology that cushions life rendered useless. Trains had stopped, phone networks were down, the power supply flickered on and off. Some commuters raced to find public phones, buried in hard-to-find corners of the train stations.[1]

We saw all of this in the seconds after the quake and realized that we had been clinging to each other like barnacles in a stormy sea. We pulled ourselves apart and slowly made for the exit. Nanako began trying to call her mom, who lived in Setagaya, a few miles from the city center, crying when she couldn't get through. On the streets as we walked the four miles toward the Foreign Correspondents' Club in Yurakucho, hundreds of office workers wearing candy-colored safety helmets crowded outside buildings, glancing nervously toward the sky. Fire engines and ambulances wailed; a siren sounded continuously from the local city office. Plumes of thick black smoke billowed from the direction of Tokyo Bay.

Salarymen (businessmen) crowded around TVs in the upmarket Ginza district, shaking their heads incredulously as they watched live reports of a huge tsunami washing away cars, houses, and towns on the Pacific coast a few hundred miles away. It was our first glimpse of the calamity that had struck the northeast, known as Tohoku, in Japan. A helicopter working for national broadcaster NHK had made it into the air from Sendai, the largest city near the epicenter, and was filming the tsunami racing inland. As the voice of the shocked reporter rose in intensity, a giant inky wave miles long began to blot out beaches, rice farms, houses, and cars. Drivers desperately trying to escape were being dragged and tossed like toys across winter rice fields and towns hugging the northeastern Pacific shoreline. Famously reserved and stoic, the Tokyoites around us struggled to articulate their astonishment. "Iyaa, taihen da" (It's awful); "I can't believe it"; "What are those people going to do?" The end of the world must look something like this, we told each other.

LUCY BIRMINGHAM, SHIBUYA, TOKYO

Just another quake, I mused after sensing a mild trembler roll through the Tokyo NHK newsroom, the national broadcaster where I'd worked since 2000. Seconds before, a cacophony of buzzing cell phones owned by the 60 or so staff in the bustling room had warned of the approaching seism. All Japanese phones are designed with an automated earthquake warning system, and the noisy phone eruption had reminded us once again that we lived on one of the most quake-prone lands on the planet.

But the massive jolt seconds later stopped us in our tracks. This one was different—not the usual back-and-forth roll but an unmistakable hard jump, a sure sign that something more intense was coming our way. I glanced out the window onto the expanse of trees and sky over Yoyogi Park and noticed a dark sweep of clouds quickly covering the patches of blue like a great, ominous warning. Below, two shirtless men were playing basketball, oblivious to the chilly March weather or undulating ground beneath them.

As the shaking intensified, the old headquarters building began to weave with an angry groan. Windows rattled hard and objects fell from shelves. As I followed everyone's lead and jumped under a desk,

I discovered a terrified colleague curled in a ball, shaking uncontrollably, eyes brimming with tears. "We'll be all right," I said, trying to calm her, but in truth my confidence was waning fast. This was by far the strongest and longest earthquake I'd ever experienced during my many years in Japan.

My thoughts immediately went to my three teenage children as a growing dread churned into nausea. As if we were caught in a bizarre time warp of terror, the seconds passed like hours while the quake's deep waves whipped like a dragon's tail below us.

When the rattling finally halted and we all emerged from under the desks, I immediately tried to reach my kids by cell phone. But the service was either jammed with calls or suspended. My text messages went unanswered, so I opted for a landline and managed to reach my Japanese in-laws before the lines overloaded with calls.

"Don't worry—we heard from Seiya and he's on his way back home," said my mother-in-law. I breathed a sigh of relief. My 19-year-old was at nearby Shibuya Station just about to step onto a train when the quake hit. Had he done so seconds earlier, it could have meant hours stuck in a packed carriage or worse. I knew my 14-year-old, Sachi, was at a friend's house not far from home. I was confident her friend's parents would take care of her.

That left Nina, my irascible 17-year-old. She was at her high school over an hour away by train, a trip that included two transfers. My heart began to pound as I considered the possibilities. Parents need great faith to live in a vast urban environment. The Greater Tokyo Area is almost the size of Connecticut, and it has a population density twice that of Bangladesh. Fortunately, Tokyo is overall a safe city for children. But in such a violent earthquake, Nina's life was in the hands of fate.

I will never forget Nina's message that arrived at 8:48 P.M.: "I'm fine. I will stay at school tonight." She had attached a photo of her and some friends lying on mats in her classroom. They were among several hundred students, teachers, and parents taking refuge there for the night. In an instant I had gone from growing despair to elated relief. The moment fueled my love and appreciation for my children. It also gave me strength to face the mounting revelations that this was becoming an unprecedented triple disaster.

After I got home in the early hours of the next morning, I went to my in-laws' house nearby and found them wide-awake, watching the unfolding news on TV. "This is quite a birthday present," said my father-in-law. Indeed, it was March 12, his eighty-fifth birthday. "I've survived World War II, the Kobe earthquake, and now one of the worst quakes in Japan's history," he said. "I sure hope this is it for a while."

For those like him who remembered the atomic bombing of Hiroshima and Nagasaki, there was worse to come. The situation at the six-reactor Fukushima Daiichi nuclear power plant in the northeast, 155 miles up the coast from Tokyo, appeared to be worsening. Operator Tokyo Electric Power Company (TEPCO) announced at 6:45 A.M. that radioactive substances may have leaked from the plant.

With limited natural resources and a reluctance to depend on foreign oil, the country's energy policy had been steered toward nuclear since the 1960s, despite being the only country in the world to experience atomic bombings. Deep government backing and close ties with the industry had set its policy in stone at that time: 30 percent nuclear with plans to increase to 50 percent by 2030. There were 55 reactors in various stages of operation and plans to build more. Serious accidents at several plants had occurred amid claims of cover-ups and falsified safety reports.

Geologists had warned of the enormous risks of building in such a seismically unstable country. But it was promoted as a safe, clean energy source, and resistance among the general population was minimal. In areas where nuclear power plants were proposed, local governments squelched the naysayers with money. The Fukushima plant, commissioned in 1971, brought jobs to the Tohoku area, known for its stagnant growth, depopulation, and historic poverty. Not a single watt of the electricity generated there was used in Fukushima. It all served Tokyo. Now the people of that prefecture were about to pay a terrible price for the deal they had made.

By Sunday morning, two days after the quake and tsunami, the news began to percolate among the foreign community that radiation carried from Fukushima in the wind and water might reach Tokyo. Rumors were circulating that embassies had recommended their citizens to get out of Tokyo, and even Japan altogether.

The exodus alarm went off on Sunday evening when the French embassy e-mailed its citizens a recommendation that they leave the Tokyo area. Other European countries quickly followed its lead. Later that week, the international schools began closing, kicking the fear factor into full gear. The US 7th Fleet moved its ships, aircraft, and personnel into open sea, away from Tokyo, because its equipment detected high radiation.

The exodus ruffled feathers in a country that can still feel closed, even xenophobic, and where distrust of foreign intentions is seldom far from the surface. There were grumbles from within many companies, domestic and international, that the Japanese staff was feeling abandoned. The term *flyjin,* a word based on the mildly derogatory term *gaijin* for foreigners, began to appear. But the mass exodus clogging roadways, airports, and train stations wasn't just a foreigner phenomenon. It involved anyone who could afford to leave, which included plenty of Japanese.

Some in the banking community flew out on privately charted jets. For most everyone else, it was long waits at immigration, in ticket lines, and at gasoline stands. The southwest region of Kansai, mainly Kyoto, Osaka, and Kobe, became the refuge of choice. Some international companies shuttered Tokyo offices to set up temporarily in Osaka, booking hotel suites and conference rooms. Hotels that had emptied just after the March 11 quake refilled with the fleeing people.

On Thursday, my children left with their father, his parents, and the dog for the Kansai region. It seemed the safest and most responsible decision. As a journalist, I felt it was my responsibility to stay in Tokyo. But saying good-bye was one of the toughest moments I've ever had. As I choked back tears, my mind raced with questions. Would this be the last time I'd see them? Was Tokyo really at risk of becoming a radioactive wasteland, as some media had reported? Was I signing away my future to cancer? Was I being a responsible parent?

Just at that moment, Nina surprised me with a heartfelt hug. Tears welled in her eyes as she then handed me a box of protective face masks. "Mom, you'll be okay. But just make sure you wear a mask," she said firmly. I winced at the thought and then promised her I'd at least try one on. "We'll get through this," she added with a smile. It was then that I realized that her generation would be the bearers of this tragedy. Would they be able to change their world?

DAVID MCNEILL, TOKYO

I arrived back in Tokyo on March 14 after three grueling days in Tohoku. Nanako's texts pulled me back from work. "It's not easy for me here with the b inside me. . . . I had nightmares and woke up . . . I'm thinking about you every second," said one message. "U have to come home asap!! Please!!!!! My tears will not stop. . . . xxx." The normally bustling, ebullient capital felt like the blood had been drained from its veins. Rolling power cuts, food and petrol rationing, the growing death toll, and the constant threat of more aftershocks and tsunamis filled the TV screens. NHK reported that workers were struggling to stop the nuclear power plant from going into meltdown. Thousands of schools closed, and factories and businesses were operating on reduced hours to save power.

Many of my foreign friends in the city had already quietly left, taking holidays from work and drifting off to Osaka or to Hong Kong, South Korea, or Thailand. But even as panic flicked on the edges of Tokyo life, all around me I saw salarymen going to work in the mornings and evenings, housewives queuing for water or milk after dawn. Throughout the worst week of the crisis, a diligent clerk from my local video rental store phoned daily to remind me that I had failed to return a DVD. It was a very Japanese crisis; things may have appeared to be falling apart, but the center would hold.

I tried to illustrate this point in an article for the *Independent,* after I parted with Nanako, who left for Osaka to escape the radiation. Exhausted and emotional after leaving her at the train, I decamped to a coffee shop in the station, where the four perfectly turned-out waitresses serenaded my entry with a singsong "Irrashaimase!" (Welcome!) and fussed over my order with typically attentive service. "Take your time," said a beaming young woman as she passed me my coffee, at which point I started crying. I pondered this admirable and mysterious ability of many Japanese to function normally as the scenery collapses around them—how people continued their daily routines, and waitresses still acted as though the most important thing in the world was my 280-yen order.

Some say that these people just fell back on routine because they didn't know any better. "Robots," said one of my friends disparagingly after I told him about the video store clerk who kept calling me. But I don't agree. Those waitresses were human beings with families who worried about radiation, too. I like to think that they stayed focused because to not do so would have been to let down others, and that invites chaos.

Would that quality be enough to revive Japan? It is not just the natural elements that have conspired against the country. In March 1945, near the end of Japan's disastrous war, US bombers dropped close to half a million incendiary bombs on sleeping Tokyo, reducing nine miles of the city to cinders and killing 100,000 people, who were "scorched, boiled and baked to death," in the words of the attack's architect, General Curtis LeMay. It was then the single largest mass killing of World War II, dwarfing even the destruction of the German city of Dresden on February 13, 1945. When the droning of bombers finally stopped on August 15, 1945, nearly 70 cities had been reduced to rubble and well over half a million people, mostly civilians, were dead.

The war killed over three million Japanese, wiped out a quarter of the nation's national wealth, and led to the confiscation of all its colonial booty, including Taiwan, South Korea, Manchukuo, North China, and Micronesia. Yet, the country engineered probably the most remarkable feat of economic regeneration in history, climbing from this humiliated postwar wreck to the world's second-largest economy in just three decades. This achievement was called the Japanese economic miracle, and there was a time, just 25 years ago, when some predicted that it would overtake the United States for the mantle of global number-one economy. Japan was down, but surely not out. But what about the impact of the disaster on ordinary lives? That's what this book is about.

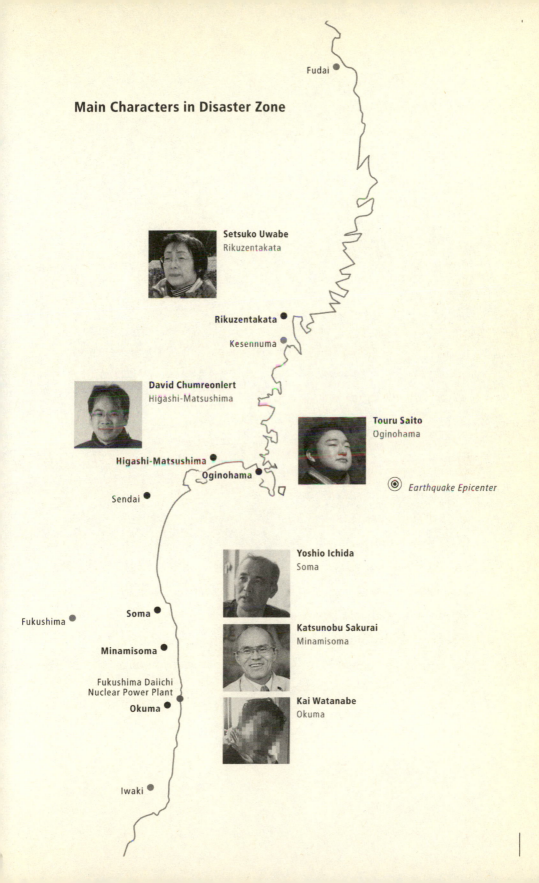

Main Characters in Disaster Zone

Fudai

Setsuko Uwabe
Rikuzentakata

Rikuzentakata

Kesennuma

David Chumreonlert
Higashi-Matsushima

Touru Saito
Oginohama

Higashi-Matsushima

Oginohama

Sendai

◉ *Earthquake Epicenter*

Yoshio Ichida
Soma

Soma

Fukushima

Minamisoma

Katsunobu Sakurai
Minamisoma

Fukushima Daiichi
Nuclear Power Plant

Okuma

Kai Watanabe
Okuma

Iwaki

ONE

The Quake[1]

Strong in the Rain
That is the kind of person
I want to be.
—*Kenji Miyazawa*

"THE WORLD IS HEAVY ON US SOMETIMES," SAYS Katsunobu Sakurai, recalling the day it almost crushed the life out of his city. The disaster began for him, as for millions of other Japanese, at work. The mayor of the coastal city of Minamisoma, Sakurai was with a group of visiting delegates on the fourth floor of the city hall when the building began to shake, gently at first, then in jerky, violent movements that seemed to go on forever. In some parts of the building, he could hear people crying. Others began pleading to the distance, to God, perhaps to the ground itself: "Tasukete!" (Help!); "Tometekure" (Please stop). Cracks opened up in the walls above his office. It was, Sakurai found, difficult to stay upright. He looked up at the ceiling of the 40-year-old building, then focused on a jug of water on the desk in front of him, catching it before it tipped over and spilled, jolted by the power of the quake. He was surprised to find himself not especially afraid. *What will be will be,* he thought.

There was nothing on the morning of March 11, 2011, that suggested it would be different to any other, or that Sakurai would become one of its unlikely heroes, his pinched, exhausted features beamed across the planet during the depths of the crisis. As he did every weekday since January 2010, when he was elected mayor, Sakurai strolled through the main entrance of the shopworn local government building, cheerfully greeting clerks before walking upstairs to his third-floor office overlooking a lattice of dense, squat housing stretching to the coast about seven miles away. When the summer sun shines, the coast is famously beautiful, anointed in the azure waters of the Pacific, attracting thousands of surfers to

Kitaizumi Beach. In the winter, the majestic mountains to the west turn snowy white, throwing the dun-colored, aging buildings in the city below into sharp relief.

A typically busy schedule lay ahead: In the morning, a meeting was scheduled with his staff, followed by a speech to hundreds of youngsters at a graduation ceremony in a local middle high school. By the end of the day, about one hundred of the city's children would be dead, some laid out in a makeshift morgue, and he would wonder, when he found time to ponder such things, if any were among the graduates he had met. After lunch, he was to meet a delegation of politicians from the Diet, the Japanese parliament, who were visiting the city. In the evening, he would see his elderly parents, both in their late 70s and struggling to manage.

A diminutive, birdlike man, Sakurai's unprepossessing appearance hid a formidable will. Locals around Minamisoma were used to seeing him jogging around the countryside, training for marathons. He had been driven into political life partly by anger. After working the land locally for a quarter of a century, he watched in despair as Fukushima Prefecture, where Minamisoma is located, licensed an industrial-waste processing plant close to his five-acre farm. All the hard work that he and other farmers had done to build up the reputation of local organic rice and vegetables, purifying the soil of chemicals, was ruined, he thought. They took the plant operator to court in the nearest big city, Sendai, in nearby Miyagi Prefecture, but after a 12-year battle, lost. Throughout the fight, Sakurai was harassed and sometimes threatened by violent Yakuza gangsters, who control much of the labor for dirty, dangerous work in such factories and who resented his attempt to block its construction.

The clash with corporate power and the sense that it had conspired with officialdom against him and other small farmers left him shell-shocked. "We weren't even allowed in the courtroom to hear the verdict," he recalls. "I was angrier than I've ever been in my life. How could people far away from us make decisions that would affect our lives so profoundly?" He decided there was no point in just using the law to fight; he had to be in government, and so he ran for office.

Sakurai's waking life was effectively parceled out among the coastal city's employees and 71,000 citizens. The days filled up with school visits, speeches, reports, and meetings with parents, farmers, and workers—an exhausting commitment to public service that left little time for his parents, with whom he shared a house. Most days he was in his office till dark, toiling beneath framed pictures of his stern-faced predecessors framed on the wall above his head.

At 55, Sakurai considered himself steady in a storm, the embodiment of his favorite poem by Kenji Miyazawa, with whom he shares an alma mater, Iwate University: "Strong in the Rain / Strong in the wind / Strong against the summer heat and snow / He is healthy and robust / Unselfish / He never loses his temper / Nor the quiet smile on his lips / That is the kind of person / I want to be." Those qualities were to be tested to the limit.

The explosive force that Mayor Sakurai and the townspeople felt at 2:46 P.M. had been released by one of Japan's most unstable faults, about 60 miles east of his office and 19 miles beneath the sea. The earth's crust is made up of eight large tectonic plates that have been moving and grinding against each other for millions of years, and the largest—the Pacific Plate—dips under the slab of rock underneath Japan's main island, Honshu. Eventually, the stress of that friction is released, but seldom as violently as on March 11. Scientists would later estimate its force at over one million kilotons of TNT—the atomic bomb dropped on Hiroshima in 1945 released fifteen kilotons. The force of the quake tugged the Pacific coastline 8 feet closer to the United States. Ancient Japanese blamed earthquakes on the angry gods. Even modern inhabitants of one of the planet's most technologically sophisticated societies sometimes wondered if they were not right.[2]

The shaking subsided. It had lasted perhaps five or six minutes. Sakurai took a deep breath to collect himself, led everyone he could find out of the building, and then began to round up his 15-member executive team. They would have to set up a temporary disaster response headquarters outside the building. As Minamisoma was a coastal city, a tsunami was very likely. The 40-year-old city building was too far from the sea to be threatened, and it had withstood the initial quake shock waves, but nobody would bet on it surviving aftershocks. People shivered in the bitter cold, but nobody wanted to risk being inside. Men and women began dialing cell phones to check on relatives, some crying when they realized the network had crashed, overwhelmed by data traffic 60 times heavier than normal. The huddle of voices around the mayor was tinged with fear, panic. Unknown to Sakurai, some of his townspeople were already dead, crushed under roofs. A 23-foot tsunami was 40 minutes away. And at the Fukushima Daiichi power plant 15 miles to the south, the power was out, detonating a chain of events that would, in a few days, turn Minamisoma's incipient disaster into an existential crisis.

Japan has an earthquake detection and warning system second to none. The nationwide online system detects tremors, calculates an earthquake's epicenter, and sends out brief warnings from more than a thousand seismographs scattered throughout the country. The system first detects evidence of P waves (for primary), which have fast, short wavelengths and do little damage. These are followed usually several seconds later by the destructive S waves (for secondary) with longer wavelengths. These snakelike seismic waves are the terrifying movements that destroy buildings and create landslides.

The system is run by the National Research Institute for Earth Science and Disaster Prevention, but it is the Japan Meteorological

Agency that sends out the earthquake warnings. It takes only seconds for a seismometer located on land, closest to the epicenter, to detect enough signals to determine if an alert is necessary. The alert is automatically issued to factories, schools, TV networks, radio stations, and mobile phones. Damaging S waves travel at about 2.5 miles per second, so they would have seriously rattled towns along the northeast coast in about 32 seconds. The S waves would have reached Tokyo, about 190 miles to the south, in approximately 90 seconds. Although the systems can only give warnings from seconds to one or two minutes before the powerful S waves hit and shaking gets serious, it can mean the difference between life and death. It can be just enough time to take cover, drive a car to the side of the road, step back from getting on an elevator, or stop medical surgery.[3]

Tsunami warnings take longer because more calculations are involved. A regional tsunami warning was made nine minutes after the March 11 quake struck. In the areas hardest hit by the tsunami, residents probably had only about 15 to 20 minutes of warning. The March 11 quake was so powerful that it triggered a tsunami that traveled all the way to the Antarctic, where it cracked an ice shelf. But despite a long history of such deadly tsunamis, years of false alarms had inured local people to the perils of coastal life all along Tohoku.

Few believed that it would be destructive, not least Mayor Sakurai as he climbed up to the rooftop of the building to observe the quake damage. As he squinted into the distance, he saw a huge cloud of dust, known as *tsuchikemuri,* rising high up in the air. "Is that a fire?" he wondered aloud. "No," he was told. "That's the tsunami hitting the shoreline." He was astonished by how big it was. "I went speechless. I realized that a huge one was approaching." Even as he spoke, the deluge was drowning the old, the young—even whole families on the coast. The waves had traveled across the sea at the speed of a jetliner—up to five hundred miles an hour[4]—then slowed and elongated as they hit the shallow coastal waters, funneled still higher by the hills that ringed the coast. Nobody could remember water reaching as far as National Highway 6, a few miles from the coast, let alone Sakurai's parents' house further inland still. Later, he learned that the water had lapped into the first floor of the house. His parents had fled. It would be days before he'd get word of whether they had survived.

ABOUT 15 MILES FROM MAYOR SAKURAI'S OFFICE in Minamisoma lay Okuma, a small town of 11,500 where Kai Watanabe was born and raised. It received the brunt of the disaster and felt the gravest global consequences. For better or worse, Okuma basked in the warm commercial glow of the local six-reactor nuclear power plant. The complex provided work to hundreds of local people, and its operator, Tokyo Electric Power Company (TEPCO), had, along with the government, pumped hundreds of millions of dollars into the surrounding economy in a process that nuclear opponents likened to drug addiction. TEPCO had put down deep local roots, rebuilding the local sports center and constructing a playground. It organized school tours to the Fukushima Daiichi plant and even screened cartoon movies for children in public halls because Okuma had no cinema.

Opposition to Japan's dash for nuclear power flared sporadically in the late 1960s and '70s as experts warned about the dangers of building plants on some of the world's most seismically unstable real estate. The experts found a largely indifferent, even hostile media during a period when the economy boomed and nuclear power was considered a badge of national pride. Some found their academic careers derailed and were treated somewhat akin to traitors.[5]

By the time Kai Watanabe was born in 1983, the intense discussions and protests sparked by the decision to start building the

Fukushima plant in 1971 had subsided. TEPCO had become one of Japan's most powerful companies, monopolizing electricity supply to Tokyo, controlling the production and distribution of a third of the nation's electricity, with "substantial" influence on politics at all levels.[6] When he graduated from high school in 2001, there was little debate in his family about where he would work. Some of his friends left for Tokyo, for college, or for the big city firms. Kai didn't even consider university. His parents ran a small Japanese restaurant and couldn't afford to send him to one. "It was seen as a perfectly natural choice for me to work at the nuclear plant," he recalls. "The plant was like the local air, and I wasn't afraid of it at all."

A 27-year-old lover of rock music, cars, and electronic gadgets, Kai was stocky and strong. His maternal grandfather had worked at the plant for over 20 years. The presence of the family patriarch, now nearly 80 and still healthy, dispelled any lingering worries about the nuclear complex. "Other people might have worried about getting cancer or leukemia or explosions or whatever, but we never discussed it once," he says.

Kai began working as a maintenance worker, checking for pressure inside pipes, opening and closing valves, measuring radiation when it was needed. He liked the work, which he felt was important. "I thought we were on a mission to provide safe power for society, for Tokyo. I was proud of that." It paid 180,000 yen (US $2,267) a month—roughly the same as a junior office clerk in Japan. A decade later, he was still being paid the same amount—plus 1,000 yen a day (US $12.60) as "bento [lunch] money."

On March 11, Kai rose as usual in the house he shared with his parents and brother, a couple of miles from the plant, and took the bus to work. In the morning, he walked around reactors one, two, and three, working from a list, checking pressure values and gauges for signs of abnormality. He went back to his office to eat lunch, then returned to work in the afternoon, finishing early at about 2:45. As he walked back to his office, he felt the building begin to sway back and forth and heard the creak and rattle of old pipes and metal on metal. *Earthquake,* he thought, his heart instantly beating faster. There had been a string of quakes in the previous weeks; some of his colleagues even joked that the big one was coming. But this was no joke. As the shaking grew stronger, he froze in his tracks. It was the strongest quake he had ever felt, and he was in the belly of one of Japan's oldest nuclear power plants. "I knew right away it was the start of something really terrible," he recalls. "I wondered if I'd live or die."

Where is the epicenter? he thought as he ran for the exit. Probably Miyagi Prefecture farther up the coast from Fukushima—there was a big one due there. He remembered reading about the earthquake four years previously in Niigata Prefecture to the west, almost directly underneath the world's largest nuclear plant—Kashiwazaki-Kariwa. It had killed eight people, burst pipes, triggered the release of radioactive steam, and forced a revision of quake preparedness at Fukushima Daiichi, including the building of its new, state-of-the-art emergency center.[7] How many would die this time? What would be the outcome of this explosive collision of state-of-the-art atomic power with primordial seismic instability? The building was now in darkness. An alarm wailed.

The exit was farthest away from the six reactors facing the Pacific, a mile or so from the sea. Once he was outside, his town, Okuma, was a few miles away. Just outside the exit on the right was the visitors' center, festooned, like many similar facilities throughout Japan, with cartoon characters. A deep 18-foot crack ran right through the silver lettering on the wall, dislodging the Chinese characters for nuclear power: 原子力. If he had seen it in a movie, he would have snorted at such contrived and far-fetched symbolism.

Maybe two hundred men were crowding around the exit, waiting to return radiation-monitoring equipment before they could get out. Line managers were counting off their staff, making sure nobody was missing: there were 6,415 people on-site, of which over 5,500 were subcontractors like Kai.

"What will happen?" some were asking. "Where do we go?"

Watching his colleagues observe procedure, Kai relaxed a little. *This isn't a country where workers drop their tools and run for the doors, to their families,* he thought. *Duty comes first.* The tsunami warning sounded as they waited. Those who couldn't get out in time went to the emergency center, waiting to be rescued. By now, Kai thought, control rods had surely been inserted into the reactor cores, shutting down the nuclear chain reaction inside. The uranium fuel was probably already cooling. Disaster had surely been averted.

But the Fukushima Daiichi disaster had already been set in train. Power to the plant's cooling system had been instantly knocked out. At 3:37, the diesel generators powering the backup would be overwhelmed by the tsunami. Even before the waves arrived, the quake may have fatally damaged pipes and the cooling system for reactor one, setting in motion the meltdown of its nuclear fuel.

During the Kashiwazaki crisis, TEPCO noticed how the river of human traffic jammed the exits as everyone tried to get out, so they introduced new rules. At Fukushima Daiichi, managers were instructed to wait outside the main gates with the authority to wave radiation checks for workers if the situation was deemed critical enough. Kai watched the plant workers line up calmly one by one and file through the gates. One or two shouted at the elderly security guards to hurry up, but that was it. If it was America or somewhere else in the West, the guards might have been punched or overrun by panicked staff, he thought. "We just behaved exactly as we always did. It was amazing to see." He left the plant and sped home to look for his family a few miles away.

THE DAY BEGINS EARLY FOR FISHERMEN, wherever they live. In Soma, a harbor town of 35,000 about 27 miles up the Tohoku coast from the Daiichi power plant, it started at 3:00 A.M. By 2:00 P.M., the working day was over for Yoshio Ichida, who lived about 100 yards from the sea where he made his living. His five-ton boat was parked in Soma Harbor, his catch of sardines, squid, and mackerel logged at the fisheries co-op, and he was sitting in the bath shaving. He would have a nap, then pick up his wife from her job at a fish-processing factory and go to see his elderly parents.

A handsome man of 53, Ichida's face is deeply tanned from life at sea. He is the third generation of his family to work the trade, and the last—his son is a schoolteacher and his daughter married and moved south. He's not unhappy—it's a hard life with an uncertain future. He and his colleagues in Soma and farther down the coast in Futaba and Minamisoma often talked about falling catches, over-fishing, global warming. There were so many worries and so many dangers affecting the sea and the fragile ecosystem that supported it that he sometimes stayed awake at night, worrying.

As he sat in the bath, he felt the entire house tremble and shake. It was an earthquake, and a strong one. Eyes wide, he watched as shelves rattled, mirrors cracked, and everything began toppling from the shelves. Water spilled over the side of the bath. Ichida immediately remembered his aging parents, but the second thought that jumped into his mind was tsunami. The ground had been shifting violently beneath Soma for millennia, and generations of his ancestors knew that the next danger, the real danger, came from the Pacific. "Everyone around here knows that quakes are followed by tsunamis."

Face half-shaved, he leapt from the bath, threw on his clothes, and ran for his car. Later he would curse himself for not taking more care with what clothes he put on as he sat shivering in the bitter cold out in the open sea, waiting for the danger to pass. He drove the short distance to his parents' place where he found the elderly pair already preparing to leave. His 75-year-old mom had been partially paralyzed after a stroke years ago. "That was strong," said his father. "I thought the house would collapse." They moved maddeningly slow. He had to bring them and his wife to safety in a local community center high above the town, then drive to the harbor and gun his boat out to sea, straight into Japan's biggest tsunami in centuries, where he could ride out the waves and save his boat. He had less than 30 minutes.

Soma's fishing cooperative was a solid two-story building that squatted on the edge of the harbor, facing the Pacific. On the first floor was a large open warehouse where the fishermen weighed and laid out their catch every morning. Manager Shoichi Abe sat every day in an open-plan office above it, eyeing the ocean out of his window as he took calls and barked orders. After decades, he knew most of the one thousand fishermen who worked there, including Ichida, who arrived on March 11 in his light pickup van and sprinted toward the water. The engines of perhaps a dozen boats were already running, churning the sea and filling the air with the smell of diesel. Men shouted and threw ropes. In the distance, trawlers raced out into the sea, but Abe knew that many had yet to leave the harbor. The boats moved in a nice, orderly line with no bumping or overtaking. He felt a burst of intense pride and admiration, especially for the first man out—his was the toughest call because he did not know what to expect. *Fishermen always help each other out,* he thought. *Even at times like this, there is a spirit of compromise.*

Two days before March 11, the area's notoriously unstable layers of subterranean faults had churned violently, triggering a strong
earthquake and a tsunami alert. Very few people along the Tohoku
coast heeded it. Most had heard the warning siren from their local
city office or the alert on national radio and TV broadcaster NHK
hundreds of times throughout their lives. The area recorded at least
three deadly tsunamis in the previous century, including a 1960 wave
that killed 142 people. Each time, the dead were buried if they could
be retrieved from the sea, higher seawalls were built, houses were
moved back from the shore, and life went on.

This time was different. Through the windows of his office, Abe
saw the water starting to crest and foam at the walled entrance to
the harbor. The water seemed to be coming from several directions
at once, but the largest wave was in the east. In the distance, he saw
another wave and it looked like the sea was churning against a cliff.
It was huge.

The destruction of Soma, like so many of the horrific moments
from March 11, is captured on amateur video. From the safety of a
hilltop overlooking the jetty, two friends filmed the arrival of the tsunami, inadvertently recording their own reactions, from jokey male
bravado to disbelief. The video shows a man walking nonchalantly
toward the ocean even as the first wave looms in the distance. "That
guy must want to die," laughs the camera operator. "The direction
of the wave is weird; I can't figure out where it's coming from," says
his friend. The inundation arrives in stages and seems to block out
the daylight, with the worst coming last. As the water reaches the
roof of the cooperative, about 45 feet above the ground, the stick-
like figure of Abe can be seen running for his life to the top of the
building, where he survives by clinging by his fingertips to a skylight.
The deluge shatters the windows of his office and carries off his chair
and table along with cars, houses, and people. For the last minute of

the video clip, the two men are almost silent, unable to take in what they're seeing.

And Ichida made the open sea. "I could feel the first tsunami arriving and the boat bouncing on top of it," he recalls. He crested two more waves, at least 45 feet high. It was like being on a fast elevator, and he instantly felt nauseous. And almost as suddenly, the sea became as eerily flat as a mirror. Behind him, men chugged for the sea before the waves broke. Fishermen all along Japan's coast know that the only way to save their boats from a powerful tsunami is to disobey the instinct to flee and drive into the waves before they hit shallow depths and crest. Once in the harbor, the latent energy stored up in the tsunami wreaks havoc on anything or anybody in its way. About one hundred of his colleagues forgot that lesson and were not so lucky.

From his boat, Ichida could see the spray from the waves crashing ashore. It rose high into the air. When he would return to port later, he would be astonished to see the devastation. The third tsunami, the most powerful, traveled over two miles inland, and the port of Soma was washed away. The water tossed boats aside like toys. The destruction included his house and the community where he lived. But the disaster was only beginning: 27 miles away, three reactors at the Fukushima Daiichi power plant were already overheating. Eventually the radioactive poison spilling out would seep into the sea, and from there into fish, plankton, and seaweed. The fishermen of Fukushima have worked the seas for centuries, before electricity, before the Meiji Restoration, perhaps before the emperor existed. What would happen now?

TWO

Tsunami

Keep running! Keep running!

—*Setsuko Uwabe*

"TSUNAMI IS COMING!" SHOUTED AN OFFICIAL AS he ran into the school gymnasium. Beneath his safety helmet, the man's face was etched with panic. David Chumreonlert was astonished to hear the news. The quake had been exceptionally strong, but should they really fear a tsunami? The elementary school was nearly two miles from the ocean, and the gym had been recently rebuilt as a tsunami-safe community evacuation site.

David stood near the gymnasium entrance doors, watching the one hundred or so children flee to safety toward the stage at the back end of the gym or crowd up the narrow stairwell to the second-floor balcony. With about 25 teachers and 20 parents who had come to pick up their children, they were able to help the elderly who had evacuated from a nearby nursing home. About 50 in total, some were strapped into wheelchairs.

After the power cut that followed the quake, the gym had become a cavernous trap. Panic whirled behind him, but David felt a strange calmness. He stared forward through the tightly closed glass doors with equal fascination and fear. *What in God's name is coming?* he thought as he glanced at his watch. It was 3:10 P.M.

A strange sound emerged. It was a low rumbling, like a fast-moving train cutting through the air, growing louder and alarmingly close. Suddenly, through the glass doors, he saw a massive mound of water outside rolling up fast between the gym and school building. Like a black monster, matted with debris, it was engulfing everything in its path. Cars in an adjacent parking lot were being swallowed or tossed like dice. "Oh Lord Jesus," he whispered as the water began to seep in from underneath the doors.

As he turned to run toward the elevated stage, a car thrust by a wave crashed into the doors. The impact broke the glass and left a gaping hole. Water gushed across the wooden floor and quickly converted the space into a rising, sucking pool. Chairs neatly lined up for the next day's graduation were washed to the sidewalls. Soon, wooden platforms and desks were bobbing in the rolling water.

David's pounding heart deafened the terrified screams of the children as the water reached the stage. The scene before him transformed into a slow-motion nightmare. Washing across, the waves hit the back wall and began dragging everyone off of the stage toward the sides of the gym, deep into the growing, freezing mass.

The water was turning David numb as he desperately clung onto the stage wall. He is not a strong swimmer and had to fight off the pulling current that had twisted him around. No longer facing the entrance doors, he could now clearly see the stage area and all those flailing in the water near him. The elderly from the nursing home were desperately trying to stay afloat and clinging to any object they could find.

An older man, with a woman clinging onto him, grabbed David's shoulder as they floated by, forming a human chain. With the extra weight, David's fingers began to slip. He grabbed onto the stage curtain nearby, while the twosome let go.

Fearing the curtain might come unhinged, and desperately wanting to help others, he jumped onto a set of wooden stairs floating by. He was pulled out to the center of the swirling water. By this time, the water had reached the level of the basketball hoop and was just inches below the second-level balcony. Another jump, some dog paddling, and he managed to grab onto the bottom of the balcony railing.

The balcony's narrow walkway was now the only safe refuge. As he tried to pull himself up out of the water, his foot slipped and landed on something hard. In the dark muddy water, he could not see the object jutting out from the wall, but it became his lifesaver. Holding onto the balcony railing with one hand and feet steady on the hard object, he could now stand about thigh-deep in the water. This gave him just enough height to survey the surrounding area nearby. He was relieved to see that most of the kids had made it to the balcony or were swimming that way, thanks to the school's rigorous swim classes. Some teachers were paddling atop floating gym mats, dragging kids and adults to safety.

Suddenly, coming toward him was one of his junior high school students treading water hard but barely managing to keep afloat. David reached out, grabbed the boy's school sweater, and pulled him safely toward the railing. A group standing on the balcony helped drag the boy from a watery death. As he turned around, David saw a woman frantically bobbing up and down, desperately clinging to her baby. As she floated by, he held her tight and was able to pull both to safety.

Frantic pleas for help reverberated above the din. "David sensei! David sensei! Help us! Help us!" It was four of his elementary school students along with one woman, all precariously clinging to a large floating desk that could capsize at any moment. "Hold on!" he shouted back, but the desk was too far to reach. "Please help us!" they cried. "Please help us!" It seemed an eternity before the desk was within reach. Finally, David was able to catch the edge with his foot and drag them to the balcony. The small drifters shimmied across the desk as the woman followed. David joined them as they crawled over the railing, collapsing into an exhausted heap, out of danger.

There was no time to rest. Out of the corner of his eye, David spotted someone trying to crawl over the far end of the balcony from the stage stairs. When he approached, he realized it was Kasahara sensei, frozen to the bone. Across from him, separated by water, were several older, terrified women desperate to reach the stairs to safety. David realized a bridge made out of the school's hollow wooden steps might enable them to get across. But each hollow step would need to be filled with water to keep it submerged and stackable.

Using hands and feet, David began the laborious, painful project. His fingers quickly turned red and raw. But just as the escape route emerged, one panicked woman jumped onto the structure, toppling it over. They were thrown into the numbing water, flailing and gasping for breath. Kasahara sensei managed to pull the woman onto the stairs as David crawled out after. He shivered uncontrollably, his wet freezing clothes clinging to him like icicles.

The water began to recede, answering his prayers for the ordeal to end. But the longest night of his life was about to begin, as he stood huddling for warmth and bracing against the quake's aftershocks among 50 other survivors in a cramped, bitter cold storage room.

The sun had been shining and the smell of spring in the air after the long Tohoku winter when David headed out from his apartment that early morning in Higashi-Matsushima. He did a quick mental check as he stepped into his car. *Friday. Right. Today is Nobiru Elementary School.* He was feeling kind of excited. It was the last day of the school year. He made sure not to forget his camera. The next day was school graduation, and he really wanted to get a few shots of the kids practicing for their big day. Like all schools in Japan, the academic year ended in March and began at the start of April.

Although the day's schedule would be a bit different from usual, he would still be with the fifth and sixth graders. They were his favorites among the four schools where he had been teaching English these past two years. The kids were smart, outgoing, and fun. It was probably a combination of good teachers. David was working with two who were equally enthusiastic about the English lessons. There was a lot of good synergy between them, and the results showed. The kids were not shy about making mistakes, like so many others kids he had taught. In fact, they were so animated and engaged that the class period would always fly by.

David, 29, was the only foreign teacher in all four schools. The number of children in the region was shrinking fast, a result of Japan's aging and declining population, especially in rural areas. Only a few native English teachers were needed. Not only was he a novelty, he was Mr. Popular. His warm personality, quick smile, and friendly demeanor put everyone at ease. Students would often ask him if he was "half," the term for half Japanese and half foreign. "No, not quite," he would answer with a wry smile, and then explain that he is American and his parents are from Thailand. They went to the United States for college and met in Texas. David and his younger brother and sister were all born in Houston. "But what's Texas? What's Houston?" they would ask. He would then ask them if they knew about NASA where they design spaceships, and surprisingly even the youngest would know, their imagination soaring as he explained further. Texas wasn't just a spot on the map of the world. It had a friendly, warm face called David sensei.

David never imagined that Japan would be his home away from home, much less a small seaside town like Higashi-Matsushima[1] along the country's northeastern coast. Nearby is famed Matsushima, a town considered one of Japan's three most scenic places and a favorite tourist destination. The bay is wide and dotted with hundreds of tiny islands topped with wind-swept pine trees. The islands, as it turned out, acted as a natural barrier against the tsunami and were largely responsible for saving most of the town.

David had applied for a job in Japan as an ALT (assistant language teacher) on a whim, at his friend's suggestion. It just seemed like a good idea to keep his options open, so he applied on the company website, not thinking all that much about it.

He had graduated from the University of Texas, Austin, in 2008 with a major in hydrogeology and had several job interviews lined up. He was expecting to work in Texas or a dry region, someplace where water is as precious as gold. But nothing came through. At the same time, the interviews with the English teaching company were going well. They wanted to expand into the Tohoku region, so they offered him a spot. Never did he imagine the water he would find.

At first, he didn't know a thing about the area, particularly about its harsh snowy winters and frequent earthquakes. He didn't mind living far from Houston. His one requirement was to have a Christian church nearby like the one he attended in Austin that was affiliated with the nondenominational Local Churches movement, also known as the Lord's Recovery. He had learned from church members that there was one in the city of Sendai about an hour away by car from Higashi-Matsushima. For David, the church is the backbone of his spirituality. It is this spirituality that kept him strong and calm throughout those harrowing tsunami moments when death perched so close at hand.

WHEN SETSUKO UWABE LOOKED OUT THE NURSERY SCHOOL WINDOW, it was mainly out of curiosity. She wondered what the townspeople of Rikuzentakata were doing after that long, frightening earthquake. In all of her 53 years living in this quake-prone region, she had never experienced such powerful, lengthy shaking. During the most violent shock waves, the nursery school's old one-story wooden building felt like it would crack open like a walnut and topple over. The heavy red slate roof was especially vulnerable.

The teachers practically crawled across the floors and tatami mats to gather up the one hundred or so children. Ranging from one to five years, the littlest ones couldn't yet walk, and the toddlers were unsteady on their feet. As the aftershocks continued, it was a struggle to change them out of their naptime pajamas into regular clothes. Mothers and relatives, faces pinched with worry and fear, soon began arriving to pick them up. The remaining children and 23 teachers huddled under protective futon padding.

A tsunami warning was being broadcast on the radio, but they had heard many similar warnings before. They could not imagine that a tsunami would reach the nursery school. It was nearly a mile from the ocean—a 15- to 20-minute walk away. Setsuko took a quick look at her watch: 3:10 P.M. The sky had turned gray and ominous, so different from the blue sky that had greeted her that morning.

As cook for the public Takada Hoikusho (Takada Nursery School), she had to prepare lunch and snacks for about 135. Today, there would be handmade dumplings, requiring lots of preparation and time. It had been 31 years since she started work as a cook with the city government, and her skills were widely admired. She was first hired to make school lunches for students at elementary and junior high schools in the area. It was 1980, the same year she had married Takuya.

Setsuko was unusual in that way. Marriage for women in Japan usually meant quitting work for life as a housewife. But still only 23 at the time, she wanted to keep active. Takuya agreed that the extra income would not hurt, as he was also working in the city government. As a career employee, he had many different jobs over the years. With their combined pay, they could eventually buy a nice house in one of Rikuzentakata's better neighborhoods. They could afford to have several kids.

At home, she could whip up delicious combinations at a moment's notice. But Takuya didn't have a big appetite, although he was very busy with work and community activities. In fact, he was quite fussy about his food. That morning for breakfast she had fixed him the usual fermented beans, grilled fish, miso soup, and rice. And like most mornings, they ate together, casually watching a TV drama on the national NHK station. Setsuko always left for work at 8:00 A.M., while her husband followed at 8:15. "Take care!" he had said as she headed out the door.

After 31 years of marriage, it was a familiar phrase often shared between them. Like most long marriages, they had had their ups and downs. But now with their son and daughter grown and living elsewhere, the couple was enjoying each other's company more and more. They had recently traveled to Hawaii and Seoul, just the two of them. "See you later," she answered warmly, never imagining what was to unfold, and how that day would forever change her life.

Looks like snow, she thought standing at the nursery school window, her eyes lowering toward the distant horizon. She gazed across the familiar rooftops toward the town's central district near the ocean. But something strange had emerged. "Is that smoke?" asked a teacher beside her. They both stared intently toward a mysterious, dusty cloud in the distance. Setsuko had never seen anything quite like it. Maybe wisps from a burning building, toppled by the earthquake, she thought. No—this had an elongated shape with enormous height. And it was quickly moving toward them.

Suddenly, like a bolt of lightning running down her spine, it became shockingly clear. "Tsunami!" she said with a gasp. "It must be a tsunami!"

Setsuko, several teachers, and the director, Keiko Kumagai, ran outside to get a better look. There in the distance they could see mounds of dark rolling water moving toward them at a terrifying speed. "Everyone, we must evacuate! We must evacuate!" bellowed Kumagai.

Setsuko and the teachers began gathering up the children. The one- and two-year-olds, ten in all, were held piggyback. The others were helped on with their shoes and then maneuvered through the door, their little hands grasped tightly by the teachers.

Once outside, the frightened group began running up a long slope toward a nearby hill. Kumagai stayed behind to help the stragglers and neighbors now joining the mass exodus to higher ground. Teachers and children from an elementary school nearby were also among the panicked, running throng. Behind them, in the central part of town near the ocean, a tsunami warning was blaring from the fire station's loudspeakers. It had started as a standard earthquake announcement and was now a terror-stricken, repetitive command. "Save yourselves! Immediately evacuate to higher ground! Save yourselves! Evacuate to higher ground!"

Setsuko looked back when the warning halted in midsentence, replaced by a buzzing sound and then deadly silence. The tsunami waves must have engulfed the fire station and areas close by. Her thoughts raced to Takuya. He worked at the city hall building across from the fire station. *Did he escape? Is he alive? How can I reach him?*

Her daughter, Kotone, would be safe in Yokohama, near Tokyo. But her son, Rigeru, was in Sendai, about two hours away by car. It was the biggest city in the region, located near the ocean and vulnerable to a tsunami. But she could not consider the tragic possibilities. Not at this moment. Not when she had to help save all of these children.

The dark seeping water was closing in on the nursery school. She could see Kumagai frantically running up the slope behind them, barely ahead of the watery grave quickly gaining ground behind her. "Hurry—we must hurry," Setsuko told the children as they struggled onward nearly out of breath, their small legs pumping. Setsuko's heart pounded. *Keep running!* she thought. *Keep running!* The children understood instinctively. Not a single one cried as they pushed forward, faster and faster.

Adrenaline pounding and choking for breath, they climbed the slope through dry rice paddy fields. As they reached a vegetable field behind a Shinto shrine, the tsunami water had managed to reach as far as the shrine.

Here in the field they finally felt safe. But looking down on the city below, Setsuko began to tremble. She could see the city hall, where Takuya worked. It had been nearly submerged by the tsunami. Only the roof remained visible, where she could see that a few had managed to escape. But she couldn't make out her husband's familiar outline.

Parents cried with relief as they discovered their children alive and safe. "We found you!" they would shout, hugging them tight. But as the hours passed and darkness descended, the freezing air dipped to about 30°F. The remaining 17 children and teachers shivered in a huddled mass as snow began to fall. The teachers knew they must find shelter.

They headed for Asunaro, a home for the mentally disabled on a hill about ten minutes away by foot. With power down throughout the area, the facility was frigid, but they were able to borrow blankets, clothes, and jackets. On hand were snacks that helped revive the hungry children. But this was not enough. Several had developed fevers. They needed to find warmth and a place to lie down and sleep.

Next door at the Kojuen nursing home, they borrowed a small room with emergency heating. They lit candles and used the light from their cell phones, but the connection remained dead. They could only wait and wonder at the fate of their loved ones and all that they had called home. Throughout the long night, parents came searching for their children. The mothers who did not come had been nurses at the prefectural hospital located in the center of town close to the sea.

They must have had a hard time moving the patients to the roof, thought Setsuko, praying for news of Takuya. A man who worked with Setsuko's husband arrived by foot the next morning after surviving the deluge on the roof of the city hall. Face pale and haggard and clothes frozen with mud, he was looking for his wife, a teacher at the nursery school. Relief flashed across his face when he found her there. But his hopes were dashed when he realized their two young children and her parents were not with her. "The tsunami probably took them as they tried to escape by car," he said, choking back tears.

"Did you see my husband?" Setsuko asked. "Was he with you on the city hall roof?" "No," he said quietly, understanding the terrible weight of his answer. But he added, "There is still hope for you."

By morning, 13 children were left with enough teachers to watch after them. Setsuko felt she could now head back toward the city and face the horrific destruction and loss that lay ahead. *I must find my husband,* she determined. *I must find him.*

UH-OH, THOUGHT TORU SAITO when his house started shaking and things began flying off the shelves. The quake's first tremor was not that strong and unusually long. He had been watching TV with his grandmother, sitting snugly under a warm *kotatsu* (heated table), eating mandarin oranges, relaxing and happy that he had finally completed his driving school course the day before.

Toru had graduated from high school on March 1 and was all set to enter Tohoku University in April, the start of the academic year in Japan. He would be majoring in engineering and wanted to learn how to design robots for nursing care used by the elderly and handicapped. With Japan's aging population, it was a growing field with lots of need.

Toru was one of the very few in his large high school to get accepted into the prestigious university in nearby Sendai based on recommendations alone. He had always worked hard to ace his grades. He was especially glad to be able to live near the university campus. The one-hour commute from his tiny fishing village of Oginohama to his high school had not been easy. He was probably one of the farthest from the school and the closest to the ocean. There was not much chance to get together with his friends outside of school. On March 11, he decided to stay at home.

A strong quake—about a magnitude 5.0—had hit the area just two days before. But this temblor was infinitely more powerful. It hit the house like a rock. He and his grandmother looked at each other and did not need to say a word. They knew they had to get out of the house and to an evacuation center.

Like everyone in Oginohama, Toru's family understood the deadly power of a tsunami. The village was within the borders of Ishinomaki City, but its location on the Oshika Peninsula left it isolated and potentially inaccessible in a natural catastrophe.

A large seawall 13 feet high had been built in the 1960s, after the 1960 Chile tsunami had damaged the coastal region. But still, everyone was cautious about going to an evacuation center after a large quake. As a child, Toru had learned about earthquakes in elementary and junior high school where they had often practiced emergency drills. Strangely, tsunami drills were not part of the curriculum, perhaps because they were confident that the seawall would protect them.

For the many elderly living in the 50 or so village households, it was difficult to flee a fast-moving tsunami. Like many rural communities throughout Japan, Oginohama was graying and shrinking. Migration to the big-city jobs had emptied small towns of young people, though it is the young, ultimately, who must carry the

Japan Tsunami History

No country has experienced more tsunami than Japan. Below are selected larger tsunami.

November 29, 684	the Great Hakuho earthquake and tsunami. The first Japanese tsunami ever to be recorded, although there were many before this. Deaths unknown.
July 9, 869	Jogan Sanriku (Sendai), 1,000 dead.
August 26, 887	Ninna Nankai tsunami, deaths unknown.
May 27, 1293	Kamakura, 23,000 dead.
August 3, 1361	Shohei Nankai, 700 dead.
September 20, 1498	Meio Nankai—Kamakura (building housing Great Buddha swept away), 30,000–40,000 dead.
February 3, 1605	Keicho Nankaido, 5,000 dead.
December 2, 1611	Sumpu (present-day Shizuoka), deaths unknown.
December 22, 1698	Seikaido Nankaido, deaths unknown.
October 28, 1707	Hoei, Shikoku, 30,000 dead.
June 15, 1737	Sanriku and Hokkaido—Kamaishi destroyed—deaths unknown.
August 29, 1741	West Hokkaido, deaths unknown.
April 4, 1771	Yaeyama Islands, 12,000 dead.
May 21, 1792	Unzen-Shimabara, Kyushu, 15,000 dead.
November 4, 5, and 7, 1854	Dai San Ansei, 80,000–100,000 dead.
November 11, 1855	Great Ansei Edo earthquake, 10,000 dead.
June 15, 1896	Meiji Sanriku, 22,000 dead.
September 1, 1923	Great Kanto earthquake, 105,385 dead.
March 3, 1933	Showa Sanriku, 3,064 dead.
December 7, 1944	Tonankai, 1,223 dead.
June 16, 1946	Nankai, 1,330 dead.
March 4, 1952	Tokachi Offshore, 33 dead.
May 23, 1960	(Japan) Valdivia (Chile) hits Sanriku, Hachinohe, 142 dead.
June 16, 1964	Niigata, 28 dead.
May 26, 1983	Sea of Japan, 104 dead.
July 12, 1993	Southwest of Hokkaido, Okushiri, 250 dead.
September 26, 2003	Tokachi Oki, 2 dead.
July 16, 2007	Niigata, at least 7 dead.
March 11, 2011	Tohoku, 19,125 dead and missing as of March 2012.

Compiled from official Japanese statistics and historical records by Geoff Tudor for *No.1 Shimbun,* the Foreign Correspondents' Club of Japan magazine.

On a 1,300-year average, a tsunami hits Japan every seven years.

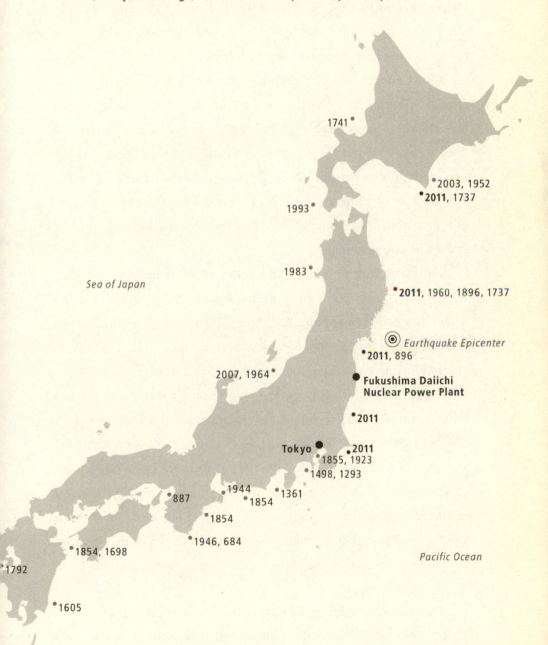

1741

2003, 1952
2011, 1737

1993

1983

Sea of Japan

2011, 1960, 1896, 1737

◉ *Earthquake Epicenter*
2011, 896

2007, 1964

● Fukushima Daiichi
Nuclear Power Plant

2011

Tokyo ●   **2011**
1855, 1923
1498, 1293

1944
887   1854   1361

1854

1946, 684

Pacific Ocean

1854, 1698

1792

1605

financial burden of supporting the elderly. Boosting the population has become a national priority. But after two decades of economic stagnation and ballooning national debt, as well as a surge in working women and limited child care, the motivation to make lots of babies is shrinking along with the numbers.

Although still sleepy from his overnight shift with the trucking company, Toru's older brother Akira quickly came downstairs and brought their grandmother to his car. Toru followed but did not think it necessary to bring supplies. It would probably be just a quick excursion. Their eldest brother, Tatsuru, was at work at a big paper company in Ishinomaki where they would have their own evacuation procedures. The brothers were not too worried about him and figured he could manage.

The boys, with their grandmother, a fit and sturdy 83-year-old, first drove to the family-run lumber factory a few minutes away where they met their parents. Started by Toru's grandfather about 40 years before, it was a small but successful company. Lumber was an unusual business in a village known for its oyster farming.

Oginohama is known, by few, as a mecca for oysters. The acclaim is transcribed in a large stone epitaph mounted near the beach. Oyster farmers along Tohoku's Sanriku coastline, like Toru's village, are credited with providing oyster fry to European farmers when an infection around 1970 wiped out the region's industry. Roughly 80 percent of oysters farmed worldwide are said to be offspring from those grown along the Sanriku Coast.[2] Stacks of neatly tied bundles of scallop shells used in the area's oyster farming are a common sight near the shoreline.

The family agreed to gather at the town hall, the designated evacuation shelter. The boys made a quick return home to get warm jackets and then sped to the shelter. Toru's mother joined them while his father headed off down toward the shoreline. As one of the village volunteer firemen, he was responsible for locking the seawall flood-gates. In this small community, able adults were assigned emergency tasks. This one could be a dangerous job.

After their grandmother was settled in the evacuation center, the brothers joined their mother in the family car, parked in the adjacent lot not far from the shore. Fitted with a small TV, it was the only place they could see the news, now that electricity was down. They learned that the quake had struck at 2:46 about 37 miles offshore and there was an enormous tsunami heading their way. The live views of the massive rolling water shot from a news helicopter were riveting. They sat transfixed at the sight of coastal towns engulfed in seconds and did not notice the water creeping up behind them.

Toru looked up briefly and something odd caught his eye. Several houses were moving. "Whoa—look at those houses!" he blurted, suddenly realizing they were being pushed by the tsunami. Within seconds, the car floorboards were covered with the seeping liquid. "We've got to get out!" yelled their mother as she tried to start the sputtering engine. By this time, they were surrounded by a jumble of other cars shifting en masse as the water climbed steadily higher.

The doors were jammed. "Keep the windows moving up and down! Don't let them get stuck!" she shouted. The windows were their only escape. The car battery still worked, and they were able to open the power-operated glass. One by one they crammed through the opening and plopped into the freezing black deluge. In an instant they were swimming for their lives.

On the other side of the submerged roadway was a thick concrete and stone wall about 20 feet high that hugged the base of a mountain. It was mottled with small holes, so they would be able to climb to the top to safety. But could they make it there? It seemed like an instant but also an eternity as they splashed through the debris-filled seawater and then scrambled to the top of the wall, their frozen fingers clutching small branches and stones on the way. Shivering under the falling snow, they watched their car being sucked vertically into the roiling mass. They knew the next wave would be coming soon.

It was clear they had to get to the area's other evacuation site, Toru's former junior high school, about a 30-minute walk away. After the tsunami wave receded, they climbed down from the wall and began walking quickly along the exposed roadway toward the junior high. They feared that another wave would soon reach them.

Fortunately a neighbor in a car came along and picked them up. They reached the evacuation site just minutes later at about 3:30 P.M. The school was already filled with students and teachers, and residents were arriving from nearby villages. Those who had originally evacuated to the town hall shelter were also coming there. Many were soaked, and everyone trembled in the cold.

The town hall was only two floors high and deemed unsafe after the tsunami waves grew steadily higher. Toru's grandmother had been rescued by a kind soul and brought to the junior high school. It was a huge relief when they found her there. *But where is Father?* Toru thought. *Is he still alive?*

Toru remembered the tsunami stories he had heard from his grandfather. The Chile tsunami was unforgettable for those who had lived through it. Because of the great distance dividing the two countries, no one in Japan had felt the earthquake that rocked the Chilean coast on May 22, 1960. The 9.5 magnitude temblor is still the largest in recorded history and produced a tsunami of 80 feet. It reached Japan seven hours later after passing through the Hawaiian Islands and destroying the Big Island city of Hilo with waves reaching 35 feet.[3]

The Japanese Meteorological Agency, responsible for tsunami prediction in Japan, had until then focused solely on the country's many local earthquakes and tsunamis. They had no experience with large tsunamis from distant locations.[4] Like a silent but deadly monster, the massive rolling waves crossed the Pacific Ocean ten thousand miles to Japan's northeastern shores with unexpected fury. Hydrodynamics, not understood at the time, had increased its size. Waves reached from 10 to 25 feet, killing at least 142 people. Damage was estimated at $50 million. Toru's grandfather would always remember seeing the drowned and bloated body of a family friend who had gone back home to retrieve a few important things.

JAPAN HAS THE WORLD'S LONGEST HISTORY of recorded tsunamis. It is little wonder that the Japanese word *tsunami* has been adopted into the lexicon of languages.[5] One of the earliest recorded was the Hakuho quake and tsunami in A.D. 684. The ancient *Chronicles of Japan*[6] includes an estimated 8.4 magnitude quake that struck Japan's southwest, near Awaji Island, and was followed by huge waves.

Japan's tsunamis have been triggered mainly by large undersea earthquakes at tectonic plate boundaries. Japan is located at the intersection of three continental plates: the Eurasian, Pacific, and Philippine Sea Plates. When the ocean floor at a plate boundary rises or falls suddenly, creating a submarine earthquake, it displaces the water above it and launches the rolling waves that will become a tsunami.[7] A quake with a magnitude exceeding 7.5 will usually produce a destructive tsunami. Tsunamis can also be caused by underwater volcanic eruptions or landslides and even a large meteorite impacting the ocean.

A tsunami is a series of ocean waves that can travel in the open sea as fast as 500 miles an hour—about the speed of a jet plane. The length from wave crest to wave crest can be 100 miles or more. Out at sea, wave heights are usually only a few feet or less, making it difficult to detect from the air or in a boat. But as tsunami waves approach shallow waters along the coast, they slow to speeds around 30 miles per hour and begin to grow in height and energy.[8] The sign of an approaching tsunami, and the signal to immediately evacuate for higher ground, is when the tsunami trough, the low point beneath the wave's crest, reaches shore and sucks the coastal water seaward. This action exposes the harbor and sea floors. The tsunami will then usually hit shore in about five minutes.

Factors that determine the height and destructiveness of a tsunami include the size and depth of the quake, the volume of displaced water, the sea floor topography, and any natural obstacles such as islands that can lessen the impact.[9] Japan's northeastern Tohoku coastline is one of the world's most tsunami-prone regions. The upper portion, called the Sanriku Coast, has been infamously dubbed the country's "tsunami coast." About 370 miles long, its breathtaking and charming vistas belie its ravaged history.

From jagged cliffs above, one can see the Pacific Ocean waves gently washing in and out of funnel-shaped ria inlets.[10] But it is here that the nightmare first unfolds. The rias are notoriously deadly. When struck by a tsunami, the funnel shape can amplify a wave's destructive power. The combination of powerful Pacific Ocean quakes and a "sawtooth" coastline shape has produced the massive height and reach of the region's tsunamis.

Until 2011, the worst tsunami in modern Japanese history was along this stretch of Sanriku coastline. The estimated magnitude of the 1896 Meiji Sanriku earthquake was 7.2, much weaker than the magnitude 9.0 quake in 2011. But the waves reached heights nearly equal at 100 feet. Approximately 27,000 were left dead or missing. The region was pounded again 37 years later in 1933 with tsunami waves described as 92 feet high caused by the magnitude 8.4 Showa Sanriku quake. This temblor and tsunami killed over 3,000 people.[11]

The size of the 2011 tsunami astonished many, but the signs were there—literally—posted on roadways all along the Sanriku coastline. Some were set high on winding, hilly roads—ancient, chilling reminders of a tsunami's long, destructive reach. Farther up, shrines could be found on sites established centuries ago, often on steep hills behind coastal towns. In all likelihood, they were built by the ancestors of the 2011 victims, knowingly far enough away from destructive tsunamis.

Blind faith in modern protective seawalls caused numerous deaths. Although the concrete walls may have helped lessen the death toll and level of destruction, most were built too low to stop the waves, and often at astronomical cost. The height of the 1960 Chile tsunami became the standard for specifications, rather than the higher 1896 Meiji Sanriku tsunami.

Poorly designated evacuation sites also added to the death toll. More than 100 sites in the three hardest-hit coastal prefectures were destroyed by the tsunami. Many fled for safety to designated temples, public schools, and community sites, only to be swept away as the tsunami waves engulfed the buildings. In Minamisanriku at least 31 of the town's 80 designated evacuation sites were inundated.[12]

Officials responsible for tsunami warnings were misguided by the initial Japan Meteorological Agency tsunami height calculation. The formula used by the agency to measure earthquakes could not accurately deal with tremors of magnitude 8 or stronger. Their first estimate, while the ground was still shaking, was 7.9. The quake that struck was 12 times larger. The resulting tsunami height prediction was about 20 feet in Miyagi Prefecture and 10 feet in Fukushima and Iwate Prefectures, much lower than actually occurred. This led to tragic delays in evacuating these areas.[13]

Power cuts and the lack of backups left many public warning systems useless. David Chumreonlert's elementary school may never have received a proper tsunami warning because of this innate problem.

The lack of tsunami education and emergency drills was also a factor. But each town and village, however, was different. In Kamaishi, leaders heeded stories from their elders of the region's 1896 and 1933 quakes and tsunamis that killed thousands. Experts warned that another huge tsunami could hit the region again within the next 30 years. Educating children about tsunamis seemed a viable disaster-prevention program. In 2005, they invited disaster social-engineering expert Toshitaka Katada and his colleagues at Gunma University Graduate School to teach emergency disaster lessons to students at the city's junior high and elementary schools.

Teachers incorporated tsunami awareness into the children's daily school lives. By 2008, the city's board of education launched an official disaster-prevention program using Professor Katada's teaching methods. Students learned about local tsunami history and even studied the physics of tsunamis in science class.[14]

When disaster struck, the students were well prepared to follow their teachers to safety. But the power of the March 11, 2:46 P.M. temblor knocked out the school's microphone system, and teachers were unable to follow the emergency procedure they had practiced.

Students at the Kamaishi East Junior High School took matters into their own hands. After gathering on the school grounds, they decided to run toward an evacuation site about half a mile away. Alarmed at the sight of the running students, teachers and children at nearby Unosumai Elementary School quickly followed, even though they had already evacuated to the school's third floor.

At the evacuation site, they met an elderly woman who told them to head for higher ground. She was convinced a huge tsunami would come after a cliff behind the site had collapsed. Everyone continued running up the road to a new location a few hundred yards away, with older students assisting the young. As they turned to look at the harbor, they could see a massive wave engulfing the city. "I looked back and saw the approaching tsunami and how houses and cars were smashing into our school," said English teacher Shin Saito. With trembling legs, Saito and the others continued on to even higher ground.[15]

The tsunami swept over the junior high school, the elementary school, and the first evacuation site and lapped within several feet of the second site. But all 212 junior high students, 350 elementary school children, and 21 teachers survived. Fourteen of the junior high students, however, lost one or both parents.[16]

A total of 1,046 people are dead or missing from Kamaishi's tragedy. But only five junior high and elementary school students perished among the 2,900 students attending the city's 14 public schools. It was clear that the city's program of disaster prevention combined with local memory lessened the tragedy that befell elsewhere.[17]

Japan's March 11, 2011, tsunami has been a tragic reminder of the need for preparedness and effective warning systems. This goes hand in hand with prediction. Mariners and scientists worldwide have been developing earthquake and tsunami prediction methodologies for centuries. And yet a prediction is only as good as its timing and delivery. The battle is convincing authorities early enough that the threat is real.[18]

In 2009 seismologists with the government's National Institute of Advanced Industrial Science and Technology (AIST) insisted that new evidence of the massive Jogan earthquake and tsunami that hit the Tohoku coastline in 869 A.D. be considered more closely in nuclear regulation. The warning fell on deaf ears.

For researcher Masanobu Shishikura, the line of destruction wielded by the Jogan tsunami over 1,100 years before had a hauntingly similar profile to the one unfolding on March 11. He and his team with AIST had done forensic research on soil layer samples in the Tohoku Sendai region. Large tsunamis striking coastal lowlands leave layers of organic debris that serve as a sedimentary record.[19] By carbon dating this material, scientists can determine the tsunami recurrence intervals.

After comparing the sedimentary record with disaster accounts chronicled in historical records, Shishikura's team concluded that megatsunamis struck the Tohoku coastline at about 500- and 1,000-year recurrence intervals. A megatsunami in the region was 100 years overdue.

Their presentation to the prefectural government was scheduled for March 23, 2011. They had even prepared warning maps for coastal residents showing areas inundated by the 869 tsunami and adducing scientific evidence that quakes tend to wield very similar lines of destruction. Those areas ended up being almost exactly the same ones hit by the 2011 tsunami. Later, he would agonize over the missed timing and many lives lost.[20]

IN THE TINY MOUNTAIN VILLAGE OF ANEYOSHI in Iwate Prefecture, the water reached a point 127 feet above sea level.[21] For those in its way, it was like being struck by a 13-story concrete wall. Villagers credit the warnings of their ancestors, carved into four-foot-high stones, with saving their lives. All had heeded the warnings and built their homes on higher ground out of reach of the waves. Residents in Sendai's Wakabayashi ward neglected history's lessons. The waves traveled a remarkable five miles inland, destroying almost 2,700 houses in one district.[22]

Water swells from the waves even crashed into Antarctica's 260-feet-thick Sulzberger ice shelf near New Zealand, causing chunks to break off and form icebergs. Some were almost as big as Manhattan. NASA scientists said they had never seen activity like this before.[23]

In the sea off Soma, Ichida listened to the splutters of information from his radio over the rhythmic wash of the now peaceful sea against his boat. The damage ashore sounded horrific, but he had to wait for the tsunami warning to be called off before he could see for himself. It was freezing beneath the cloudless moonlit sky, and his boat had no drinking water or food. Throughout the long night of March 11–12, he tried to keep his mind off what he would find. Unlike many out on the ocean, at least he knew there was someone waiting for him. Before the phone network crashed, he had received a text message from his wife: "Family is safe. No House."

Just after noon on March 12, he could stand it no more and gunned his boat for shore. As Soma Harbor loomed into view, chunks of debris bumped the hull, and he had to slow to a crawl. He struggled to believe what he was seeing. It looked like photographs he had seen of Japanese cities after the war. He did not even know where to pull into the harbor because all of the landmarks were gone, except for the concrete skeleton of the fishing cooperative. The van he had parked almost 24 hours before had, of course, been swept away. The ground had liquefied in places, and the sea banks, which had taken 30 years to build, were in pieces.

On the walk to his two-story house, less than a mile away, he swapped news with neighbors he met. About 60 of his friends and colleagues were dead or missing, swept away by the waves. Everyone had a horror story. Nothing was left of his house but the concrete base. The entire neighborhood had vanished. He was not a crying man and was used to hard work and the ups and down of life at sea. There would be no tears—only resignation. *We can reconstruct this,* he thought. *Our ancestors have been through worse.* But as the sun fell in the sky on March 12, he and his colleagues began to hear bad news from the nuclear power plant up the coast that his ancestors could never have imagined.

THREE

Close the Gate

The floodgate will protect us for sure.

—*Masafumi Saito*

❝I'VE GOT TO CLOSE THE SEAWALL FLOODGATES," SAID Toru's father urgently. As a veteran member of the village volunteer firefighters, it was Masafumi Saito's duty to lock the gates that protected the seaside residents from tsunamis. Toru, his brother Akira, his grandmother, and his parents had gathered at their family lumberyard, located only about 65 feet from the sea. What should they do next? The earthquake was probably the biggest Masafumi could remember. A tsunami was surely on its way. *Maybe like the one last year—only a few feet high,* he thought. *The floodgate will protect us for sure.*

"Go to the evacuation center. I'll meet you there," he said firmly, knowing his family would heed his word. Masafumi was not a large man, but at 47, he was still strong as an ox. Like the local fishermen, he had been bred on the area's harsh nature. His powerful hands, rough from years of working with lumber, could easily handle the floodgate's three locking methods, a combination of an iron wheel and iron doors that slid and swung closed. "Take care of yourself," said Toru's mother as they parted.

Because Masafumi was not a fisherman, he was one of the few volunteers available to lock the gates. As he approached the compact harbor, he could see the villagers' fishing boats already out at sea. At the helm were young men and sons of elder fishermen who had the muscle to safely maneuver their boats over the tsunami waves. The boats could easily be damaged if left moored at the harbor.

Masafumi immediately set to work. Three other volunteer firemen were supposed to join him as part of the emergency procedure, but they were nowhere to be found. *Looks like the other guys couldn't make it,* he thought. *Probably held up 'cause of that big quake.* As he glanced down at the shoreline, he spotted a buoy attached with a net full of oysters that had been washed ashore. *Maybe the tsunami already came,* he wondered hopefully.

The concrete seawall was about 13 feet high and 427 feet long. Closing the four floodgates was a familiar job and usually took him about 20 minutes. As he reached the third gate and looked again toward the bay, he noticed a strange movement. "What is that?" he mumbled aloud. The water was whipping a cluster of debris in a wide circular motion. Round and round it went, gaining speed. He strained to get a better look. *Oh no,* he thought. *It's a whirlpool.*

He raised his eyes and quickly focused on the ocean beyond. A massive, high wall of water was rolling over the breakwater protecting the bay and speeding his way. He stood for a second, awed by its formidable power. The sight was surreal. But then a surge of adrenaline raced through him, powered by terror. *Run! Run!* he told himself. As he bolted inland, the tsunami wave crashed over the floodgate. A dark hideous mass, it rushed ashore, nearly lapping at his feet. As he raced forward, he turned his head briefly and caught sight of an elderly man flailing, then disappearing into the sucking wave.

The cemetery. The cemetery, Masafumi repeated in his mind. Ironically, the hillside graveyard was the highest and safest point nearby. Like many countryside cemeteries in Japan, it was in an elevated spot, placed out of respect for the dead, and often adjacent to a Buddhist temple.

The vista was breathtaking on a beautiful day. But here and now, the sight below was terrifying. Masafumi managed to reach the gravestones unharmed. Breathless and jolted by his brush with death, he huddled with several elderly fishermen who had seen off their boats just minutes before.

As snow began to fall, the shivering group headed for a Shinto shrine nearby that sat at the top of a long mountainside stairwell. Not only did it offer high shelter, it had candles they could use in case they got stuck there overnight. But after a brief time there, Masafumi decided to brave the continuing rush and receding of tsunami waves and find his family. *I'll head for the nearby town hall evacuation site,* he decided. *Surely they were safe there.*

Minutes later when he reached the town hall, he was shocked to find that it had been inundated by the tsunami. Where was his family? *They must be safe at the junior high school,* he thought, while fighting off a growing dread. Another designated evacuation site, it was about 30 minutes away by foot, on a good day.

It was slow going along the roadway, now muddy and full of debris. About every ten minutes, he would scramble up the hillsides as each tsunami wave reached toward him and then receded. Along the way, he helped elderly neighbors also struggling toward the evacuation site.

A biting, rank smell suddenly filled his nostrils. It was propane gas. The tsunami had broken open gas tanks located near the shore. The vapor had permeated the air like an invisible toxic blanket. *Will there be an explosion?* he wondered fearfully. It would turn into an inferno if spread to the mountainsides. They would not have the manpower or access to control it, especially now.

It was about 7:00 P.M. and dark as he approached the school parking lot. He was surprised not to see the family car. A frightening thought flashed through his mind. *Did my family make it?* As he glanced across the lot, he spotted their familiar outlines. They were laying out what appeared to be wet clothes. He was shocked to see them. The clothes they wore were caked in mud, and they were shivering hard. *My God—what happened?* he thought.

When they told him about their harrowing escape, they agreed it was a miracle that everyone had survived and was reunited. When Masafumi revealed his terrifying moments at the floodgate, he felt an agonizing conflict of sadness and rage. Later, they would learn that the tsunami height reached 32 feet. Fortunately among Oginohama's 130 villagers, only three were swept away, but all homes and properties were destroyed. "The gate didn't protect the village," he says painfully. "We trusted it too much."

JAPAN HAS GONE TO GREAT LENGTHS to protect itself from tsunamis, typhoons, and damaging waves. Breakwaters constructed offshore that deflect strong ocean waves and seawalls and floodgates built at shorelines are a common site along the country's 22,000-mile coastline. Japan has about 12 massive antitsunami seawalls located in harbors nationwide and smaller seawalls as high as 40 feet along more than 40 percent of the country's shoreline.[1]

But in northeast Tohoku along the Pacific Coastline during the March 11 tsunami, nearly all of the concrete and metallic defenses proved to be useless or only partial deterrents. Seawall heights were often far below the tsunami peak. The thick, reinforced concrete walls were overpowered by the pounding waves. Many of the seawalls dropped in height after the undersea quake, and massive tremors that followed shifted the seabed located below. In some cases, the seawall deflected the waves toward towns nearby, compounding the destruction there. Some walls even trapped the seawater and prevented residents from escaping.

The tsunami's path of destruction stretched from town to town along Tohoku's upper northeastern Sanriku Coast, with shattered seawalls testaments to the power of the sea. Massive chunks of concrete lay buried in dirt and sand, along with the extraordinary expense.

Jagged corners jutted out like graves for the missing. Jumbled in between sat mounds of useless wave-breaker concrete tetrapods, tossed along the beaches like a game of jacks. In one town, the Beatles' song "Yesterday" became the forlorn 5:00 P.M. chime heard daily over the community loudspeaker.

The 2011 tsunami is now reshaping shoreline defense. Along National Route 45, which hugs the coastline near the village of Fudai, there was a harbor filled with hundreds of brand-new dolos, concrete blocks to be used for a future seawall project. Each structure weighed up to 20 tons and was engineered to dissipate a wave's energy, not just block it. The region was clearly fortifying itself against more great and violent threats from the sea. Sadly, it was too late for the many towns destroyed on March 11.

IN FUDAI ON MARCH 11, fireman Takehiro Furuma[2] was relaxing at home on his day off, playing with his seven-year-old son, when the quake struck at 2:46 P.M. After the first jolt, he quickly turned on the TV news. A massive earthquake had hit the region, and a large tsunami was expected to follow. As a 14-year member of Fudai's firefighting team, he was well trained to handle the area's capricious quakes and sea swells. He knew he had to get to the firehouse right away. But the shaking emerging from this temblor was so violent that he, his son, and his wife could not move.

Fudai had been destroyed by quake-induced tsunamis in the past. Now, the three thousand residents were shielded behind a breakwater 51 feet tall and a huge concrete and steel floodgate extending between two mountainsides. When the late mayor Kotoku Wamura pushed the project through in the 1970s, it was considered a classic, rural boondoggle costing the taxpayers an astronomical $42.5 million (¥3.5 billion).[3]

Wamura served over 40 years in office between World War II and 1987 and became a powerful political figure. He was known for his close ties with local construction companies and the region's political personalities.

"Go to the evacuation site!" Takehiro shouted to his wife and son after a pause in the shaking as he bolted out of his house to the firehouse nearby. It was 2:50 P.M. when he joined the men at the floodgate's remote controls set in a small room at the back of the firehouse. Video cameras monitoring the gate 24/7 were showing all systems go as four of the six gates began to automatically close at 2:55 P.M. The two end gates were built on roadways and needed to be checked onsite and cleared of people and traffic. Takehiro and a fellow fireman immediately went to check the one closest to the elementary and junior high schools and closed it easily with the push of a button.

Takehiro then headed off to an emergency briefing with the mayor and village heads as his colleague went to check the other roadway gate. "All systems go!" the colleague relayed by radio transceiver at 3:03 P.M. as he headed back to the firehouse. But just two minutes later, at 3:05 P.M., the gate suddenly stopped, leaving a gaping hole for the tsunami to rush through. The problem seemed to be an electrical outage possibly due to the force of the quake. Someone had to go back to the gate and close it, despite the oncoming danger.

Four firemen were chosen, two full-timers and two volunteers. Closing the gate by handwheel would take literally three hours. Fortunately they were able to fire up the emergency backup gas engine by 3:25. The heaving iron gate began to close again, but not quickly enough.

"Get the hell out of there!" they heard over the walkie-talkies at 3:27 P.M. "The tsunami is right on your tail! Get out!" The four

Seawall and Tsunami Heights by Town

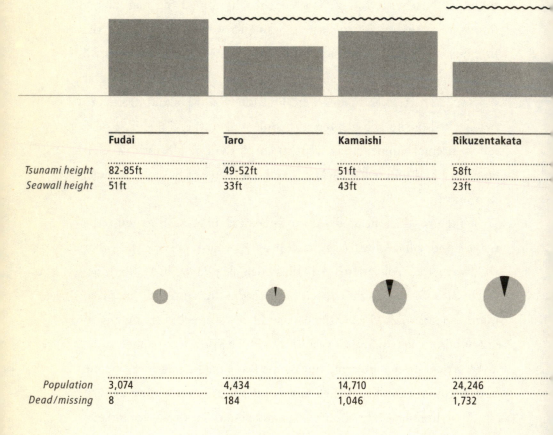

	Fudai	Taro	Kamaishi	Rikuzentakata
Tsunami height	82-85ft	49-52ft	51ft	58ft
Seawall height	51ft	33ft	43ft	23ft

	Fudai	Taro	Kamaishi	Rikuzentakata
Population	3,074	4,434	14,710	24,246
Dead/missing	8	184	1,046	1,732

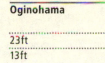

Oginohama

23ft
13ft

Ishinomaki

28ft
20ft

Higashi-Matsushima

34ft
20ft

Minamisoma

39ft
20ft

130
3

162,822
3,819

43,153
1,120

71,494
947

firemen could hear a low thundering noise, but the massive floodgate blocked them from seeing the raging water heading their way. One of the new volunteer firemen was fiddling with the gate controls. "Leave it!" the others shouted at him. They scrambled to their vehicles and sped off, barely in time.

The monster tsunami poured over the entire floodgate and raced just behind them for more than 300 yards, destroying everything in its path. It finally slowed down to a dribble after reaching a fork in a shallow central river. Not a single home had been harmed. A photo from one of the security cameras shows the moment the tsunami breached the floodgate at 3:28 P.M. The firemen had less than 60 seconds to escape a watery death.

The floodgate saved everyone who was in Fudai Village at the time. One fisherman was swept away when he went to check his boat at the port. Seven who were shopping at a neighboring town were also carried away, along with most of that town. For some in Fudai, their faith in the floodgate was so great that they chose to stay at home rather than go to an evacuation site. They were lucky. This was a tragic mistake for thousands in other towns and cities who overestimated the strength of their seawalls.

Mayor Wamura's controversial Fudai floodgate was driven not by the grubby imperatives of Japan's voracious construction lobby but by the searing memory of the 1933 tsunami that pulverized the northeast. "When I saw bodies being dug up from the piles of earth, I did not know what to say. I had no words," Wamura wrote in his autobiography, *A 40-Year Fight against Poverty*.[4] He described the disaster he witnessed in his 20s and the lessons he learned. The persuasive mayor, who passed away in 1997 at the ripe age of 88, is now lauded as a local hero. After March 11, hundreds of local people made the pilgrimage to his grave with flowers and silent prayers of thanks.

His old friend, Michishita Shigetada, 97, has been gathering do-nations to build a memorial monument in his honor. "If Wamura didn't build that floodgate, I'd be dead by now and someone would be lighting an incense stick in my memory," he says with a laugh.[5] The nonagenarian calls himself half senile, but his steady stride and clear vision defy the claim. Shigetada is among a formidable genera-tion of elderly Japanese, like Wamura, who survived the worst of times and lived long to tell the tale.

"Wamura was someone who looked gentle and calm on the out-side but was very strong willed when he decided to get something done," he says. "He wanted to build the floodgate because he was very worried about the children." With the village's narrow forma-tion, set like a valley between two mountains, the schools had to be built near the ocean, the only wide land available. If a tsunami struck during school hours, the children would be swept away.

The planned memorial monument would not be just a show of appreciation to Wamura, but also a reminder for the village chil-dren about the power of tsunamis. "I want the younger generation to always remember to *fear* tsunamis," explains Shigetada. "We can never prevent tsunamis. We can only try to lessen the damage."

JUST LIKE FUDAI, THE FISHING VILLAGE of Taro also built a tower-ing seawall. But unlike Fudai, the structure in Taro was not fortified between two mountainsides and built, like most, along an unob-structed shoreline. And yet, residents were so confident it would protect them that on March 11, they ran up to the top to view the oncoming tsunami, only to be quickly swept away. As the rolling waves surged over, neighborhoods below were trapped between the seawall and a mountain behind.

The structure was one of Japan's largest, dubbed the "Great Wall of China." Stretching 1.5 miles and 34 feet high, the bulwark loomed over the 4,400 inhabitants, blocking views of the ocean beyond. Like many towns along the Sanriku Coast, Taro had a history of destructive tsunamis, and residents were taught about the dangers.

The town of Kamaishi had the world's deepest breakwater that was listed in the *Guinness World Records*. The 207-foot-deep structure, standing nearly 20 feet above water and one mile long, was finished in 2009 after 31 years of construction. The cost: $1.6 billion.[6] Unfortunately this structure also failed to protect.

Justification for the extraordinary expense goes back to Kamaishi's importance as the birthplace of Japan's modern steel industry in the late 1850s. Nippon Steel became its biggest employer until the early 1970s when it moved to central Japan, where auto industry production was booming.[7]

Construction on the breakwater began in 1978 with plans to attract new international maritime-related companies after the loss of its steel giant. But like many ports along the Tohoku coastline, business interest was fading along with population numbers and economic stability. Huge government subsidies followed, solidifying cozy ties between government and local business. Kamaishi's breakwater morphed into a colossal concrete folly. That it didn't manage to stand up to the tsunami and protect the city's 40,000 residents has added insult to injury. Records show that 1,046 residents were killed or remain missing.[8] Worse: some believe that the breakwater deflected waves onto neighboring towns and increased the destruction there.[9]

In Rikuzentakata, it was tourism that determined the town's seawall height. The wall was built only 21 feet high to keep from being an eyesore, below the height of the town's 70,000 magnificent pine trees planted near the shoreline. The trees were a main attraction for a steady stream of tourists. Since 1927, the town had been selected as one of Japan's most scenic spots.

But the March 11 tsunami waves reaching 58 feet easily rolled over the wall. Left behind: one lone pine tree, almost one million tons of debris, and a town "wiped off the map."[10] More than 1,700 residents were swept to their deaths or still listed as missing[11] from the town's pretsunami population of 24,000.

Rikuzentakata's pine trees surely helped lessen the tsunami's impact. The residents called the one pine that survived "miraculous," and it became a symbol of resilience. Salinized, rotting roots, though, have made death imminent, despite wide efforts to save it. The tree has become a symbol of rebirth with germinated seedlings and tree branches successfully grafted into a red pine.

For Setsuko, Rikuzentakata's famous beachside pine trees will always remain a fond memory. When she was young, her family would go and picnic under the shade of the tall trees and bathe in the cooling sea to escape the summer heat. She also went there for school excursions and a fun Girl Scout camping trip. She and her husband, Takuya, had their first date there. She still vividly remembers sitting with him under the stars, talking about their future dreams.

ALONG THE SANRIKU COAST, beyond seawalls and floodgates, are tsunami defenses in all shapes and sizes that bear witness to Japanese ingenuity: elaborate elevated platforms and shelters, berms (artificial high ground), spiral towers emulating the biblical Tower of Babel, and the natural defenses of adjacent higher ground, sculpted forests (mainly spruce), and islands.

A natural barrier of 250 tiny pine-topped islands protected the town of Matsushima from the tsunami's full impact. The famous bay of islands is designated as one of Japan's three most scenic views. The town's seawall was only about six feet high, but the town was left largely intact. Only 16 died among the 15,000 residents. David Chumreonlert had a sensational view of the islands every week from a high point on a nearby mountain island where he taught a small group of elementary school children.

Also in view in the distance, without any protective island barrier, was a school where he taught that was destroyed by the tsunami. It was in the town of Higashi-Matsushima, not far from where David lived. There, 33-foot tsunami waves easily rolled over the 20-foot seawall, killing more than 1,100 of the 43,000 residents.

This example and others have inspired engineers, architects, and town planners to propose defense alternatives. The tsunami-ravaged town of Natori is considering a proposal by architect Keiichiro Sako that would re-create Matsushima's survival success within a safe Utopian environment. Called Tohoku Sky Village, it would comprise little communities set on man-made elevated islands built on the destroyed land. Almost like modern-day fortified castle towns, Sako's proposed oval-shaped structures would have a width of 656 feet and a height of 65 feet, assuring safety from large tsunamis. Natori did not have a seawall and was hit by 30-foot waves during the 2011 tsunami.

The isles would be bolted into the bedrock with exterior walls reinforced with concrete. Each would be divided based on usage. Most would be multiuse residential spaces. Others would be commercial and include factories and facilities for the area's agriculture and fisheries sectors. The ambitious plan even includes an indoor marina to protect local fishing boats. Sako estimates that the ¥20 billion (US $252 million) cost for each island would be less than moving entire communities to higher ground.[12]

As advisor to the city of Kamaishi, veteran architect Toyo Ito has developed a proposal that incorporates landscape elements and existing seawalls. The design features a sloping man-made berm that provides tsunami protection for terraced apartments built within. Each building would contain 20 units with ocean views. A landscaped seaside park reaching 43 feet in height and lined with cherry trees would provide protection for a series of traditional, A-framed, thatched-roof communal dwellings, each with 16 residential units. An area much like Fisherman's Wharf in San Francisco would also attract visitors and boost the local economy. The plan would include increasing the height of Kamaishi's nearby seawalls from 13 to 20 feet.[13] Ito's proposal would cost much less than Sako's but provide fewer residential units and not include schools, commercial-use structures, or public facilities.

The aim of most local governments is to move residents to higher ground, a safe distance from a potentially lethal tsunami. But uprooting and scattering communities would incur great expense, both financial and emotional. Close community bonds, a vital source of support in Tohoku, could easily be broken. Much of the region's economy is based on fishing and life near the sea.

Setsuko wants to stay in Rikuzentakata near family and lifelong friends, despite the lack of shopping, transportation, and other conveniences. She cannot imagine living anywhere else. Toru's father, Masafumi, is sure that the residents from ruined towns and villages like Oginohama along the Sanriku Coast will slowly return and rebuild their homes and communities. "The sea can be cruel at times, but it's always abundant and giving," he says. "History will repeat itself."

For residents in communities near the Fukushima power plant accident, scattered by both the tsunami and radioactive dangers, returning may never be possible. How will history recount their desperate stories?

FOUR

Meltdown

It was a journey to hell without a map.

—*Kai Watanabe*

E VEN AS KAI SPRINTED FOR HOME, THE WORKPLACE HE left behind was skidding toward the planet's worst nuclear crisis in 25 years. The quake's shock waves ripped pipes from the walls, toppled lockers, and buckled roads at the 864-acre plant. The ten technicians and single shift supervisor in the main control room near reactor one would later describe that the shaking was so hard that some fell to their hands and knees.[1] "Stay calm!" shouted unit-one superintendent Masatoshi Fukura. The tsunami was 49 minutes away.

Initially, Fukura and his boss, plant manager Masao Yoshida, believed that Daiichi's defensive engineering had worked. The instant the tremors struck, control rods were automatically inserted into the plant's three working reactors to shut down nuclear fission, a process known as "scram." Nuclear power complexes basically operate on the same principle as coal- or oil-burning plants: water is heated into pressurized steam, which is used to turn turbines and generate electricity. The difference is in how the water is heated: by a process called nuclear fission. It begins by splitting an atom into two by, for example, bombarding the isotope uranium 235 with a free neutron. The splitting of uranium fuel, formed into long rods, releases intense heat and, without water to cool it, would overheat and melt, releasing potentially deadly radiation. Water, then, is vital to nuclear power plants. Quakes and accidents can shut down electricity grids feeding power to water pumps and other instruments, which is why all plants must have backup power.

Reactors four, five, and six were off-line for maintenance. The engineers quickly learned that the quake had cut the plant off from the main electricity grid, leaving no power to pump water to the nuclear core and carry off the heat, but 13 backup diesel generators would keep emergency water pumps running till power was restored. The generators were considered more than enough to keep the plant's juices flowing.

Nine months previously, there had been an unintentional dry run for this scenario. On June 17, 2010, power to water pumps for reactor two failed. Fukushima Prefecture's former governor, Eisaku Sato, was one of several observers who repeatedly asked what would happen if the backup generators also stopped working. It was essentially a rhetorical question. Even after nuclear fission ends, fuel rods give off intense heat. Fuel that is not cooled can heat up to 5,000°F. This heat boils off all of the water surrounding the fuel rods, exposing them to air. In a worst-case scenario, without water the fuel can melt through steel, concrete, and anything else in its way.

TEPCO did not allow for the possibility that those 13 generators could stop working. They should have learned from another rehearsal four years prior. In July 2007, a 6.8 earthquake struck 12 miles from Kashiwazaki-Kariwa, by some measure the world's largest nuclear power plant. In the seconds after the tremors began, pipes burst, drums of nuclear waste toppled, and monitors stopped working. A fire in an electrical transformer burned unattended for over two hours, and 1,200 liters of contaminated water sloshed into the sea. TEPCO subsequently admitted that the damage to the seven-reactor, 8,200-megawatt complex "extended to the interior of a reactor building" and that a small amount of radioactive water had also escaped from reactor one. Many fundamental weaknesses and failures in safety procedures came to light afterward. "The inadequate response by Tepco to the unfolding events at Fukushima Daiichi should not have been a surprise to anyone," concluded a damning March 2012 report by the American Nuclear Society.[2]

Between 2002 and 2006, 21 separate problems at the Fukushima plant were reported. The whistle-blowers, which included employees at the plant, bypassed both TEPCO and Japan's Nuclear and Industrial Safety Agency (NISA), the main regulatory body, because they feared being fired. The information was ignored. Sato would later describe how whistle-blowers were treated like "state enemies."[3]

Sato was but one of a large cast of extras in the nuclear drama who had predicted catastrophe. Seismologists cited Japan's most powerful modern seismic event, the 1707 Hoei quake, which triggered a huge tsunami that washed through much of Shizuoka Prefecture, south of Tokyo, and would surely overwhelm the defenses of most nuclear plants if repeated. In 1933, 92-foot waves demolished the northeast coastlines of Aomori, Iwate, and Miyagi Prefectures, close to the Daiichi plant. A 125-foot wave had crashed ashore in 1896. There was evidence of at least seven magnitude 9 quakes along the north and northeast Pacific coast in the past 3,500 years.[4]

A week before March 11, TEPCO and two other utilities persuaded the government to soften the wording of a report warning that a massive tsunami could hit the northeast region.[5] The government's Earthquake Research Committee subsequently altered the draft report to say that "further study" was needed because data was "insufficient." Three days before the disaster, TEPCO itself had released a three-page briefing paper indicating the need to assess the 40-year-old plant's tsunami disaster risk. The paper cited in-house computer simulations and other studies suggesting that a tsunami as high as 33 feet could hit the nuclear complex.[6] Given its five-decade track record of ignoring seismic data, it is unlikely TEPCO would have acted on it, says Hiroyuki Kawai, a corporate lawyer who would a year later lead one of the biggest lawsuits in history against the company.

Kawai was one of many voices outside the mainstream questioning not only the logic of building 54 commercial reactors in a country that experiences 20 percent of the world's magnitude 6 earthquakes, but also the logic of building them so close together: the Fukushima Daiichi and Daini complexes are about seven miles apart. Like all precision machinery, nuclear power plants are highly susceptible to water and shock.[7] Major seismic events deliver both.

Many of Japan's reactors were planned or online before modern seismology uncovered hitherto undetected fault lines in coastal areas. Scientists uncovered several particularly vulnerable power plants, notably the five-reactor Hamaoka plant in Shizuoka Prefecture, about 113 miles from Tokyo, which sits almost on the boundary of two restless tectonic plates: the Eurasian and the Philippine Sea. Kashiwazaki, too, sits on a major fault.

The studies forced the authorities to accept that a magnitude 8 quake could strike the region at any time—government forecasts have predicted an 87 percent chance of a powerful quake near Hamaoka in the next 30 years. The possible consequences for Tokyo are chilling: a Fukushima-scale accident would "signal the collapse of Japan as we now know it," warned seismologist Katsuhiko Ishibashi.[8]

Inside the earthquake-proof bunker at the Daiichi plant on March 11, manager Yoshida and his deputies began to take stock of what had happened. It was just after 3:00 P.M. Kai and thousands of workers had been allowed to leave to check on their families. Convinced that the crisis had been contained, the remaining men inside the bunker paid little attention to the sea. The tsunami struck the plant with waves of 43 to 49 feet after washing over a mile-long breakwater and the 19-foot seawall. The waves were twice as tall as the highest wave predicted. Water flooded the basements of the turbine buildings about 450 feet from the sea, on the ocean side of the reactors and lower, shorting out electric switching units and disabling 12 of the

13 emergency generators and then backup batteries, the last line of defense.⁹ The control room was pitched into darkness.

Flashlights winked on one by one. Dread filled the half light. There was no power to operate or even monitor what was happening to the reactors, or to measure radiation. Henceforth, estimating water levels inside the reactors would simply be guesswork. Four and a half hours later, the water in reactor one had dropped below the bottom fuel, exposing the fuel core. Fuel melt had begun. Many experts suspect that even before the tsunami arrived, the quake may have fatally damaged the cooling system of reactor one.¹⁰ Just over 15 hours after the power loss, the fuel melted through the reactor's pressure vessel. Reactors two and three were not far behind. Even worse, there was no plan for what to do next because nobody in TEPCO had ever predicted total loss of power at a nuclear plant.

In Tokyo, the government's top spokesman, Yukio Edano, quickly appeared on TV. "At this moment, no problem with the reactor itself has been reported," he told the nation in his first press conference after the quake. Nuclear disaster had been declared at 3:42 P.M., and at 4:45 P.M. TEPCO told the government it had lost control of the plant, meaning that it had suffered a complete loss of power.

A few hours later, Prime Minister Naoto Kan chaired the first crisis meeting on the unfolding nuclear drama, pulling together officials from TEPCO, NISA, the Nuclear Safety Commission, and economy minister Banri Kaieda. Remarkably, no minutes were taken of that meeting, but they were later reconstructed from interviews. "Meltdown is a possibility, right?" asked one of the men in the room, the first mention of a word that would be blacked out for weeks by Edano and TEPCO. When NISA spokesman Koichiro Nakamura let slip the following day that meltdown was a "possibility"—meaning core fuel melt inside at least one of the reactors—he was removed from his post.¹¹

The government immediately sent emergency generators loaded onto trucks to the Daiichi plant, but when they arrived, they found the electrical panel board they needed to reroute power flooded. The key aim was clear: somehow get water back into the reactors to cool the overheating fuel. Workers were sent out to scavenge car batteries to keep monitoring instruments working. At 7:00 P.M., Kan publicly declared a nuclear emergency, ordering the precautionary evacuation of everyone within 2 miles of the plant and telling thousands of others within a 6-mile zone to stay indoors. The following day, the evacuation zone was widened to 6 miles, then 12 miles.

In the small restaurant they had run for decades in Okuma, Kai's parents grabbed what they could on the morning of March 12 and fled inland to the small town of Tamura, then Iwaki City, about 21 miles south of the plant. They were never to return. As they drove away on a bus, a half-remembered word began reverberating, unwanted, in Kai's mind: "Pripyat." It was the name of the Ukrainian town evacuated in the aftermath of the Chernobyl disaster that is still, 25 years later, a nuclear ghost town. Kai had a premonition that they might never come back home.

Okuma emptied within hours, along with Tomioka, Futaba, and Namie, small, tidy little towns surrounded by picturesque fields, hills, and harbors where people had farmed and fished for generations. Pets and farm animals were left behind to become feral or die. The off-site nuclear emergency center in Futaba, built to handle such evacuations, was useless, with no electricity or phones or even filtering systems to keep out the radiation. Officials in Namie heard about the evacuation on public radio. Watching the disaster unfold on NHK, Kai could not believe his eyes.

At Futaba Hospital, a few miles from his home, chaos reigned in the hours and days after the ground began shaking. The tremor had been met with screams and whimpers; then came rumors that a tsunami was on its way. Rumors also swirled about what was happening inside the power plant next door. Phone networks were overwhelmed by incoming calls. Remarkably, in a country with a plan for everything, there were no contingency plans for an emergency of this kind. In the following 48 hours, the hospital staff would get about 200 of the 435 patients out before transportation workers began refusing to drive near the Daiichi plant. It was days before everyone could be evacuated. Of the patients at the hospital, 21 died in the immediate aftermath of the disaster, some strapped into wheelchairs on buses en route to evacuation centers. There would be nearly 600 similar deaths in the coming days and weeks.[12] Futaba's mayor, Katsutaka Idokawa, would later call the disaster a meltdown of Japan itself.

Back at the plant, engineers began to realize that they could die from either a blast or exposure to the deadly toxins inside the reactors. Radioactive steam and hydrogen were accumulating in reactor one and seeping into the control room. Late on March 11, they faced a classic nuclear dilemma: vent the steam into the atmosphere or watch the reactor's containment vessel explode, releasing much worse radiation. Without electricity, however, the vents had to be opened by hand, a possibly fatal task.

After hours of confusing and incomplete information from NISA and TEPCO, and wondering why the vents remained closed, Prime Minister Kan flew the 155 miles by helicopter from Tokyo to the plant and ordered TEPCO engineers into the reactor building. It was, most believed, a suicide mission: the temperature inside the building was over 100°F, radioactivity levels near the vents were at near lethal levels, and the men would have to work in pitch darkness amid a string of aftershocks. Working from old blueprints in 17-minute bursts—the maximum time they could endure without absorbing fatal radiation—the masked and suited men cranked open rusting valves. It was not enough. In the afternoon of March 12, the first hydrogen explosion ripped the reactor building. The managers inside their windowless bunker felt the blast first, then watched in horror the images on commercial TV. Radioactivity began seeping into the bunker and all around the plant, the toxic plumes spreading widely due to the brisk spring winds.

The heaviest contamination was blown north and northwest over Minamisoma and pristine farming land in Fukushima, one of Japan's key food baskets. In the following three days as more explosions struck the plant, a plume hit the mountains that ring Minamisoma and rained down on the town of Namie and the mountain village of Iitate, about 25 miles northwest of the stricken plant.

Fearing panic, NISA bureaucrats withheld data from a hugely expensive radiation tracking system called SPEEDI (System for Prediction of Environment Emergency Dose Information) that showed the direction of the plume, though they released it to the US military in Japan.[13] Thousands of evacuees from the towns and villages around the plant fled into what would prove to be the most contaminated areas. The mayor of Namie would later call the decision to withhold the data "akin to murder."[14] "It was a crime," agrees Mayor Sakurai. "The government didn't protect its own citizens." Kan claimed that the data was sent to the prime minister's building but intercepted by NISA bureaucrats before it reached him.[15]

Immediately after the March 12 hydrogen explosion, Sakurai, whose town stood a mere 12 miles away, watched Edano try to reassure the public at a televised press conference. "Even though the No. 1 reactor building is damaged, the containment vessel is undamaged," the chief cabinet secretary told reporters. "In fact, the outside monitors show that the [radiation] dose rate is declining, so the cooling of the reactor is proceeding."[16] Any suggestion that the accident would reach Chernobyl levels was, he said, "out of the question."[17]

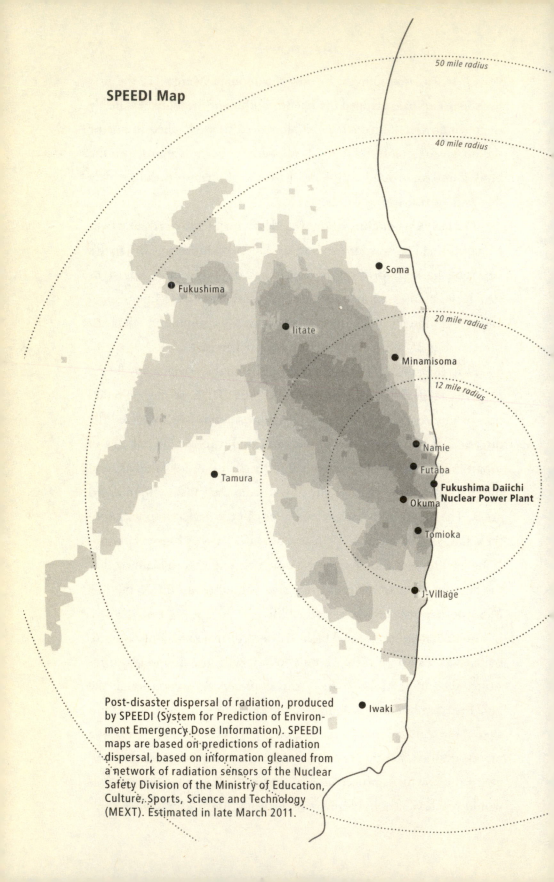

SPEEDI Map

50 mile radius

40 mile radius

20 mile radius

12 mile radius

● Soma

● Iitate

● Minamisoma

● Fukushima

● Namie

● Futaba

Fukushima Daiichi Nuclear Power Plant

● Okuma

● Tomioka

● Tamura

● J-Village

● Iwaki

Post-disaster dispersal of radiation, produced by SPEEDI (System for Prediction of Environment Emergency Dose Information). SPEEDI maps are based on predictions of radiation dispersal, based on information gleaned from a network of radiation sensors of the Nuclear Safety Division of the Ministry of Education, Culture, Sports, Science and Technology (MEXT). Estimated in late March 2011.

The world's worst nuclear disaster, which had left behind a 1,100-mile wasteland that still remains almost devoid of people a quarter of a century later, would be invoked frequently as the crisis wore on. Soon, experts in Japan's media would also predict high radiation in Fukushima for decades to come.

The playing down of the crisis was not unique to Japan. After Chernobyl, the Soviet authorities famously hid the severity of the meltdown and radioactive release, then harassed or even imprisoned those who questioned the official version of what happened. The Pennsylvania government also withheld information during the 1979 Three Mile Island partial meltdown. President George W. Bush was accused of manipulating Environmental Protection Agency data on airborne toxicity from the 9/11 attacks in New York City.

The citizens of Minamisoma, on the border of the exclusion zone, did not believe Edano. Once they saw the explosion on TV, they immediately began to leave. The exodus started on March 12 and turned into a flood by March 15, creating a traffic jam outside the city government building. Cars inched by the mayor's office, the faces of children pressed up against steamed-up windows. Day after day, hundreds of people crowded the reception area below Sakurai's office, demanding information and help. Men cornered the mayor when they saw him walking through the first floor. "What the hell are you doing!" some shouted. "Tell us what's going on, you asshole." He knew little more than his accusers. There was no communication from the government or TEPCO. Calls went unanswered. It was a week before anyone from the central government arrived and 22 days before TEPCO finally told Minamisoma about the drama unfolding at the power plant a dozen or so miles away.

Gasoline was the most common demand as people started to flee, but it had to be rationed because after the first reactor explosion, delivery trucks began to stay away. City officials were sent to man pumps at the local gasoline stand. The day after the first explosion, a tanker driver called from Koriyama, about 32 miles away, and said he was not going any farther because he was terrified of the radiation. Sakurai's staff had to go themselves and get the truck full of vital fuel. It was among the first ominous signs that deliveries to the city would stop coming. Food trucks and other utilities also ceased.

An exodus of city and medical workers from the city's biggest hospital began. How were they to cope with the sick and old? Who would retrieve the bodies still scattered around the city's coastal communities? On March 14, journalists from Japan's big daily newspapers and TV companies covering Minamisoma suddenly disappeared, meaning that on-the-ground news from the most vulnerable large city in the nuclear crisis would vanish for weeks. They would not return for over a month. Some of the city workers had started to peel off, too. "My city was melting down," Sakurai says.

In the middle of the chaos came the worst task of all: the exhausted mayor had to visit the makeshift morgue in the local agricultural college. The bloated bodies of men he had farmed with, friends of his family for years, were laid out on the ground. "There were just no words for what I felt," he recalls. But there could be no question of deserting his post or even going to look for his parents. Duty to the citizens who had elected him came first. Only later would he find out their fate. At night, he would curl up for a few hours in a room behind his office, wrapped in a blanket. Before dawn, his eyes would open and he would wonder what horrors the new day would bring.

As the sun rose in the sky on a crisp Monday morning, March 14, another blast tore apart the concrete building around reactor three, which contained large amounts of plutonium among its cocktail of lethal poisons. The explosion stopped the water that was cooling reactor two, worsening its already critical state. Daiichi was now a mess of tangled metal and rubble. Radiation in some parts of the complex was 1,000 millisieverts an hour, enough to quickly induce radiation sickness. Even with filtering systems laboring to keep out the contamination, levels of radioactivity inside the control room leapt twelvefold. The engineers stayed, rooted by loyalty to their company, duty . . . and fear. "Even if they ran, where could they go?" asks Kai. Some workers off-site, from Minamisoma and Iwaki, even tried to navigate buckled, flooded roads back to the plant, their huge fear tramped down by the weight of solidarity and empathy.

That night, NISA canceled its usual hourly press briefings, an ominous indicator of the chaos behind the bureaucratic wall erected to manage the crisis. In the bunker at the Daiichi plant, engineers were openly contemplating a once unthinkable scenario: three reactors completely out of control and spewing a vast toxic cloud toward Tokyo, the world's most populated metropolis. Many were digesting the most terrifying news of all: 1,500 fuel rods in the reactor four building, normally covered 16 feet below water, had boiled dry, raising the specter of a nuclear fission chain reaction and contamination far worse than a reactor meltdown.

As Tokyo slept, plant manager Yoshida, wearing his blue boilersuit, pulled together his staff. A tough man with a reputation for independent thought and plain speaking, Yoshida was characteristically blunt. "Go home," he said. "We've done what we can here."[18] Afterward, debate would rage about whether he and TEPCO had ordered a full or partial retreat, leaving some workers behind to stop the plant from sliding completely out of control. Some of the engineers had tears in their eyes. They thought that Yoshida was opting to die. Like the captain of an atomic-age *Titanic,* he would go down with his ship.

Trying to put together what happened at the refugee center in Iwaki, Kai feared the worst. The center was crowded with people from his town who sat fearful and transfixed in front of the TV, but few were as qualified as he was to imagine how bad this could get. A professor from the elite University of Tokyo was saying that there was no cause for alarm, but it was obvious that there was no water in the reactors and that the fuel was melting. *Why are they saying it isn't melting down?* he wondered. *We're looking at a Chernobyl-type situation—maybe worse.* Eventually, he thought, the evacuation area could stretch to over 60 miles, or perhaps nearly 125 miles.

And where is the company's management? he wondered. TEPCO president Masataka Shimizu had disappeared from public view amid rumors that he was in the hospital, overwhelmed with stress, or that he had attempted to commit suicide. It would be a month before Shimizu even came to Fukushima to apologize to furious locals, by which time his reputation was in tatters and his career ruined. Kai was not that surprised. In the Daiichi plant hierarchy, he and his friends considered the TEPCO management deskbound plodders, graduates of Japan's elite universities—men with too much head and no heart, unlike blue-collar grunts like Kai who kept the plant running. They had no experience of crisis.

Still, as soon as he saw the explosion on March 12, he began waiting for a call from his company, asking him to save the plant and clean up the mess. And when it came, he would not hesitate to say yes. Instinctively right leaning, Kai often thought in military terms, recalling wartime stories he had read in manga comics and seen on TV. "I thought of myself in the mode of kamikaze," he recalls, referring to the young men who strapped themselves to flying bombs in a doomed attempt to defend Japan from US invasion at the end of World War II. "They risked their lives for their families, for their villagers and people they didn't know. That's what I thought of: protecting people. It wasn't defending my country; it was people's faces I saw in my mind. Even if we have to give up our lives, we would allow people from our hometowns to someday return home."

The call came later, about a week after the crisis began. "We have to go back," said Kai's manager. He used a military term: "final battle orders." Some people refused the request, saying they had children or they feared that the worry would drive their wives crazy. Kai was single, and his family never spoke about the possibility of his returning—tacit approval in his mind. There was no disguising the danger. There could be more explosions and even worse radioactivity. There was just no way of telling.

FIVE

The Emperor Speaks

Please do not abandon hope.

—*Emperor Akihito*

NOT A CLASSICALLY BEAUTIFUL CAPITAL, TOKYO'S attractiveness comes from its exuberance and energy: the majestic chaos of its ticker-tape highways and railway lines, the Sturm and Drang of its vast pulsing modernity and impossibly large population. In the days after March 11, this energy drained out of the city, leaving it pallid and wrung out.

In Shibuya, Tokyo's buzzing, youthful heart, the giant neon signs and TV screens that tower over the normally clogged main intersection went blank to save power. The river of traffic and crowds thinned, bars emptied, and shops closed early. Many of the people gathered around Shibuya Station, a key hub to western Tokyo and beyond, carried cases and overnight bags, preparing to flee the city.

Taxi drivers complained that they could not get gasoline, even after lining up for hours. Power was voluntarily cut across the city. Stores ran out of water, milk, and even umbrellas amid rumors that the rain was contaminated with Fukushima's payload—reminiscent of the "black rain" that had fallen after the atomic bombing of Hiroshima. It was, some said, like stories they'd heard about the war: rationing, queues, the dark talk of calamity from the skies, which had arrived almost exactly 66 years previously on March 10, 1945, with the firebombing of the city by US wartime bombers, incinerating 100,000 people, leaving the city in ruins, and creating five million refugees.[1]

This time, the danger from the sky was radiation, which was detected in the city center, as hospitals, universities, and mass media outlets began independently monitoring the air. Millions looked to the government and the media for a clear picture of what appeared to be a looming disaster: the meltdown of a nuclear power plant 155 miles away and the showering of fallout over 35 million people. But the government seemed to be playing down the dangers, and information in the media was garbled. It was the stuff of science fiction novels, or a dispatch from the pages of *Godzilla,* the story of the fictional monster who played on atomic-bombing memories, rising from Tokyo Bay to destroy with a blast of radiation from his mouth.

The government said a Chernobyl-style catastrophe was unlikely. Modern nuclear plants are built better, and the Fukushima reactors had been shut down since the earthquake on March 11. Government spokesman Yukio Edano appeared twice daily on TV, his ubiquitous factory suit conveying hands-on reassurance, to relay dispatches that were not based on scientific information that the government itself had gathered from the plant, but which often came directly from officials in TEPCO headquarters. "Radiation levels around the nuclear plant are not at levels to cause an immediate health risk," he said on March 16. "If someone were to stay in the area 24 hours a day, 365 days a year, they might suffer health problems. But the radiation is not high enough to affect the human body over several hours or even a few days."[2]

His reassurances did not quell the sense of impending doom in the capital or the steady stream of rumors and apocalyptic headlines. "Radioactivity in Greater Tokyo at 100 Times Normal Levels," blared the *Sponichi* tabloid on March 15. France sent planes to evacuate its citizens. The embassies of Iraq, Bahrain, and Angola announced that they were closing. Panama and Austria evacuated their ambassadors to Kyoto. Two days later, the British Embassy, until then a steady, sober voice in the storm, advised on its website that British citizens should "consider leaving Tokyo and the area north of the capital."[3] The US Embassy, meanwhile, was secretly planning the evacuation of 90,000 of its citizens.[4] Officials from the US military and State Department were better informed about radiation levels and risks than ordinary Japanese citizens.

Throughout the crisis, foreign reporters were struck by how Tokyoites continued to soldier on even as panic flicked at the edges of life. There was no descent into lawlessness and savagery, or even rioting of the kind that worsened the aftermath of natural disasters elsewhere. The city's army of dark-suited workers marched to work every morning. Those few who asked for time off were told that the government had pronounced the capital safe and that if they left they would not have a job when they returned. Dread lurked below the surface of life, but unless Japan's corporations, government, and broadcasters pushed the panic button, it seemed that the city would soldier on.

Millions watched the battle to save the Fukushima plant on television. The nuclear fuel inside its reactors was dangerously overheated, and there was no power to pump water, so soldiers from the Self-Defense Forces were ordered to the site. The troops dropped water from tungsten-reinforced planes flown 300 feet above the reactors. It was clear that the effort was not working, as the wind blew the water harmlessly off course. On TV screens, it looked like the men were fighting a house fire with spray from a toy gun. American Global Hawk surveillance drones sent over the Daiichi plant confirmed that radiation levels were unchanged. Probably the most sophisticated radiation measuring equipment in the country was aboard the US Seventh Fleet, stationed near Tokyo. So when the supercarrier USS *Ronald Reagan* moved away from the coastline on March 14, alarm bells rang in editors' offices around the world. But most Japanese never heard the news.

Few ordinary people realized the gravity of the situation. While many suspected that the reactors in the Fukushima Daiichi plant were more critically damaged than the government or TEPCO had acknowledged, it would not be until May that they learned that the fuel inside units one, two, and three had already completely melted down and that fuel rods inside unit four were in a state of near criticality. At least two explosions on March 12–14 had sent clouds of radioactive contamination over the nearby countryside and sea.

Behind the scenes, a desperate battle was under way to prevent an even worse catastrophe. The prime minister and the cabinet spokesman both feared what Edano later called a "demonic chain reaction," meaning Fukushima Daiichi could trigger meltdowns elsewhere. That included the possibility of a catastrophic failure at the Fukushima Daini plant, about seven miles south of Daiichi, and at the Tokai power plants along the same coastline. The situation at Fukushima Daini was declared a level 3 incident on March 18, 2011, meaning dangerous levels of radiation had escaped. The head of plant operations at Daini, Naohiro Masuda, would later admit that the plant, too, was near meltdown. "If that happened, it was only logical to conclude that we would also lose Tokyo itself," Edano said afterward.[5] TEPCO's president, Masataka Shimizu, who had disappeared from public view, repeatedly called the prime minister's office with garbled messages, apparently demanding that the Daiichi plant staff should be allowed to evacuate.

At about 3:00 A.M. on March 15, Prime Minister Naoto Kan was woken from a nap in a back room of the prime minister's residence in Kasumigaseki, central Tokyo.[6] Exhausted and still wearing his rumpled utility suit, he sat up and went to meet the government's economy minister, Banri Kaieda, who told him startling news: TEPCO officials and seven hundred workers were preparing to abandon and flee the Daiichi complex. If true, the plant's six reactors and seven nuclear fuel pools would be left to spiral out of control. The Fukushima Daini plant would almost certainly have to be abandoned, too. "A total of 16 nuclear fuel pools," thought Kan. The most apocalyptic scenarios suddenly looked very real. A serious radioactive fire could send cesium and other toxins to greater Tokyo, forcing the evacuation of 35 million people.

Kan pulled his advisors together for a quick conference, then ordered his driver to take him from the government buildings to the utility's headquarters, a charmless, bunker-like building about a mile away. Storming into a crowded conference room, he shouted, "Why on earth are there so many people here? The important decisions should be made by five or six people." He then trained his fire at Chairman Tsunehiko Katsumata and the company's top executives. There was "absolutely no way" the workers could pull out, he said. Abandoning the plant would create a disaster "twice or three times worse than Chernobyl." If people had to die, they would die. Workers over 60 did not have to worry about getting cancer 20 years down the road. Kan himself was 65 and would go to the Daiichi plant if he had to, he said. It was no longer a question of just Japan, he shouted. The disaster would affect the world. He ended with a very Japanese appeal, tapping into a rich cultural vein of obligation and duty: "Onegaishimasu; ganbatte kudasai" (I'm asking you; give it your all). It was not a request. There would be no retreat.

The prime minister left behind a room full of chastened TEPCO executives. Later they would deny there was any plan to completely abandon Daiichi. Sakae Muto, a vice president, later told an independent panel of 30 professors, journalists, and lawyers who spent six months investigating the Fukushima accident that the utility only intended to withdraw "some of the workers." Kan's opponents said that he had isolated himself inside the prime minister's office by refusing to trust qualified bureaucrats and technicians who knew far more about what was happening than he, then needlessly panicked. Kan insisted he understood the desperate situation inside the plant only too well. "Reactor 1 had exploded on March 12th; Reactor 3 on March 14th; Reactor 2 looked very dangerous. In that situation, it's not surprising that some people would conclude that there was nothing else they could do and that it was time to pull out."[7] He would insist that the issue was one of life or death, not just for TEPCO workers but for many others.

As the battle to save the Daiichi plant raged, a rumor began circulating in Tokyo: the emperor had left his palace in the city for Kyoto, Japan's ancient capital about 230 miles away. The rumor was passed around bars, traded between taxi drivers, but it was never mentioned in the mainstream media, where such conversations are taboo. It was untrue, but even in a country where the former emperor Hirohito had been removed from his wartime perch as living god six decades ago, talk of being abandoned by his son Emperor Akihito was disturbing to some, a sign of the city's perilous psychological health.[8]

The Imperial Palace dominates Tokyo's center. The key link to Japan's mythologized past, the reclusive family inside claims an unbroken heritage dating back 2,600 years, surviving revolution, succession crises, earthquakes, fascism, US firebombing, and postwar constitutional change. Even during the war, when Tokyo was incinerated, legend says that Emperor Hirohito stayed while his then 11-year-old son and successor Akihito was evacuated.

So it was a shock to many when the now 77-year-old, silver-haired Akihito abruptly appeared on network TV on March 16, looking somber in a dark suit. His five-minute prerecorded speech, from a room decorated with traditional screens, inevitably rekindled memories of World War II, brought to a close by Hirohito, who famously said in a prerecorded radio speech on August 15, 1945, that Japan would have to "endure the unendurable." As he spoke, about 40 percent of the area of 66 Japanese cities lay in ruins and over

150,000 people had been incinerated in Hiroshima and Nagasaki. The home of tens of millions had been destroyed; millions more were evacuated. For elderly Japanese, even in 2011 those memories were still alive and vivid.

Emperor Akihito, in his first crisis broadcast to the nation, said that his "heart deeply ached" watching the tragedy unfolding. "It is my hope that many can be saved. Please do not abandon hope," he pleaded, adding that he was "deeply concerned" about the nuclear crisis. "I hope from the bottom of my heart that the people will, hand in hand, treat each other with compassion and overcome these difficult times," he said. And then with a brief bow to the nation, he was gone.[9]

It was the first direct, simultaneous imperial media address since the horrific summer of 1945. Emperor Akihito used plain, polite Japanese that was easily followed in millions of homes across the country. In the northeast, children were shushed and TVs were turned up in refugee centers so he could be heard. Commuters inside Tokyo's hub stations gathered around to watch. The sight of the high-voiced, diminutive royal slowly reading his script struck some outside Japan as an unlikely source of emotional strength, but his historic status as the father of the nation made him, for some, a collective rallying point.

Later the aging emperor and empress knelt on the floor with distressed refugees from the northeast, a symbolically important gesture. TV pictures beamed across the country focused on how the emperor sat at eye level with the people he met.

For Kai Watanabe, the speech brought mixed feelings. Like most young Japanese, he views the emperor as a remote almost spectral figure, irrelevant to his life. "He's a symbol, and of course I'm thankful to him for his speech. It was a good thing he did, a symbolic gesture. But no matter what he says, his words will never bring back the people who died, or return the Fukushima nuclear plant to its original state, or bring back the houses that have been washed away." The elderly, soldiers, and perhaps government bureaucrats might have been moved by the televised dispatch from the Imperial Palace. "But for us, the younger generation," Kai says, "it didn't mean much."

Stranded with his wife and parents in a converted junior high school in Soma with no TV or radio, fisherman Ichida did not hear the emperor's speech. He did not even know that there had been a hydrogen explosion at the plant on March 12 and missed the now famous grainy TV pictures showing debris and smoke from the blast shooting into the air. The rumors arrived soon after, carried by cell phone and whispers. One by one, families began to disappear from the temporary refugee center, pulling away quietly to Niigata or Yamagata Prefectures. "We stupidly stayed there in silence. Why couldn't they have told us?" He heard a rumor that Soma's mayor had run away. When they met later, the mayor explained that he had gone to Tokyo to get relief supplies.

About one hundred miles away, near Tokyo, Ichida's daughter began crying as she watched the drama unfold. What on earth was happening up north? She began calling her dad and pleaded with him to move farther away from the power plant, but he would not budge. His wife obeyed him. She did not want to risk the lives of his elderly disabled parents by moving. "I was giving up by then," he recalls. "I didn't mind dying there." It was difficult to explain, even a year later; he just felt that he had to accept his destiny. His daughter cried and pleaded every night, then threatened to come up and get her parents and grandparents. The two fought every day on the phone for hours, often till after midnight. In the end, Ichida's wife and family persuaded her not to risk coming to Soma if she was planning to have children. Although he knew little about radiation, he was scientifically correct: at 53, he was less vulnerable to radiation than his daughter or her offspring.

Conservative views of the emperor, found mainly among the elderly, verge on the mystical. The tide of the nuclear disaster was turned not by the emperor, however, but by ordinary workers. Prime Minister Kan's confrontation with TEPCO, beamed by closed-circuit cameras to the plant, may have been a turning point. The government ordered elite Tokyo firefighters into the Daiichi complex to cool down its dangerous fuel pools. The men were all over 40, chosen because they had already had children. Some kept the dispatch order to the front line of the nuclear crisis secret from their own families. Those who spoke of it did so in the tersest language. "I'll be back safely; trust me," one man texted to his wife. "Please be the savior of Japan," she said in return. Television commentators noted the odd, unwanted echoes of 1945, when young kamikaze pilots sacrificed themselves in a doomed attempt to save Japan from US invasion.

Their mission was to bring water to the overheating fuel pools, but this meant connecting a series of long, very heavy hoses from the sea to the reactor building in just 60 minutes. With debris scattered across the plant, fire trucks could not be used, so the hoses had to be carried by hand over a quarter mile, as radiation meters bleeped alarmingly. It was exhausting, dangerous work. With a backup team constantly shouting out radiation levels, the men laid the pipes, then fled the plant. A handwritten sign on their truck said, "We are determined to return home."

As the firefighting squad drove back to Tokyo, radiation levels at the plant began to drop and continued falling through the night. It was the first positive development since the crisis began, and TEPCO moved quickly, calling into battle the reserve army of workers and subcontractors waiting outside the exclusion zone. Their job was to keep water going to the reactors and to clean up from the mess left behind by March 11. Radiation had fallen sufficiently for younger men to be considered. The boss of Kai's company called him in the refugee center, asking him to come back. He said yes immediately.

In mid-March, the weather is still cool in Fukushima. Working in hazmat suits, layers of gloves, and full face masks, Kai and thousands of other men fanned out across the complex, clearing away the tons of debris left behind by the tsunami and reactor explosions and laying miles of pipes. In parts of the complex, radiation was so high that the men could only work in two-minute spurts, even after the government temporarily and controversially raised the maximum annual limit for workers from 200 to 250 millisieverts. The United Nations Scientific Committee on the Effects of Atomic Radiation said that 30 millisieverts a year could cause cancer.[10] The twisted and tangled mess of steel and concrete that used to be reactor three was especially feared, throwing off radiation so high that it would induce sickness in minutes. A year later, it would still be too dangerous to approach.

In April, after weeks of expert opinion scoffing at Chernobyl comparisons, the Japanese government officially raised the Fukushima crisis to International Nuclear and Radiological Event Scale (INES) level 7—the same as the 1986 Ukraine disaster. As radiation levels inside the plant continued to fall and more water was pumped into the fuel pools, the mood inside the control bunker lightened. The dreaded words "Oshimai da" (It's the end) were no longer heard. Nuclear workers believed that they might live to talk about what happened. But conditions stayed harsh. As spring turned into summer, it became unbearably hot. During the day, the air turned a sticky 90 degrees—and even hotter inside the protective suits and masks. "Men were dropping like flies in the heat," says Kai. "But they just had to keep going."

As he journeyed masked and suited up through the deserted villages and towns where he had grown up, Kai felt like he was carrying the *hinomaru* (Japanese flag) into battle to save Japan. He remembered a tune that he would hum to himself in the long days ahead. Later he would say that the song "Sakimori no Uta" (An Ode by an Ancient Japanese Coastguard), from a poem about lonely troops dispatched by the ancient emperor to protect Japan's distant marine outposts, encapsulated his and his colleagues' struggle.

Please tell me / If all living things in this world / Are destined to live limited lives / Is the sea mortal? Is a mountain mortal? I sometimes ponder upon the miseries of human lives / Upon the agony of life / Upon the pains of illness / Upon the misery of dying / And upon my present self.

Work went on round the clock. The men survived on instant noodles, rice balls, and bottled water and slept in shifts anywhere they could find a spot on the floor inside the quake-and-tsunami-proof control bunker, surrounded by the sound of filters loudly laboring to keep out the radiation. Inside the bunker, they joked that they were in the safest place in the world. Radiation here was much lower than in the exclusion zone. There wasn't even a TV to see what was happening outside.

Finding workers became increasingly difficult, even with rumors that unqualified men were getting 100,000 yen (US$1,260) a day. Men signed on for the work but then fought with their wives and families in the evenings and quit. Recruiters began driving around Kannai in Yokohama and Sanya in Tokyo, areas where the unemployed and the homeless congregated in the shadow of gleaming glass and concrete spires. Yakuza gangsters muscled in on the lucrative temporary contracts.[11] Kai's salary with the subcontractor stayed the same—about $2,400 a month—plus a small daily bonus. He knew that the workers who brought the Chernobyl plant under control had received monthly payments and medals of honor. He joked darkly that the most he'd probably get was a TEPCO towel and a commemorative ball-point pen.

When his shift ended, he was driven back to J-Village, a former soccer training ground on the outskirts of the exclusion zone that had been converted into the nuclear crisis center. From the bus window, he could see the abandoned fields, villages, and towns where he grew up. Spindly weeds had started to sprout from crevices and walls, reclaiming once neat roads and gardens. Bodies still lay unrecovered in some areas. Occasionally an abandoned dog or farm animal would stroll across the road, blocking the bus. He was not the type to cry easily, but sometimes he could feel tears welling up during these drives.

He and his colleagues rarely talked about what they did, but if the discussion of motivation came up, Kai's answer was the same. The plant was broken, and it was his duty to go and fix it. But for him, there was something more. "The plant was in the town where I was born. I wanted to go back there, to decontaminate it." But the bravado was mixed with fear that he was giving up the chance of a normal life. "Let's say I tell a woman about my past, that I've absorbed all this radiation and may get sick or father children that are deformed so we shouldn't have children. Is there a woman out there who would accept me? And if I have to work, will she understand it? Neither of us would be happy in a situation like that. So I think it's best for me to stay single. I don't talk about this stuff with my colleagues. It's taboo. We think about the impact on ourselves but not on our unborn children."

The first verdict on the government and TEPCO's handling of the crisis came almost a year later. Speaking on behalf of an independent panel of 30 professors, journalists, and lawyers who spent six months probing the accident, lawyer Akihisa Shiozaki characterized the overall response to the crisis as "crude, reactionary, but lucky." Nobody was "even remotely prepared," he said. "A small group of politicians out of panic and distrust became increasingly involved in the technical management of the on-site accident. Unfortunately, this political intervention was rarely effective, triggered confusion, and at times was extremely dangerous." But Kan's move to confront TEPCO and stop the pullout was critical. "The worst would have happened. . . . Fukushima would totally have gone out of control."[12]

That verdict was disputed by the official Diet (Japanese parliament) report released in July 2012. The Fukushima Nuclear Accident Independent Investigation Commission said that it was "difficult to conclude" that TEPCO was preparing to completely abandon the Daiichi plant. The report accepted that TEPCO and the nuclear regulators had failed at multiple points to communicate to the government the situation at the Daiichi plant, creating an "atmosphere of distrust" among TEPCO, the regulators, and the Prime Minister's Office. But it found no evidence of a plan for full withdrawal at the TEPCO head office, instead blaming Kan for disrupting the attempt to end the crisis at the plant, the "true frontline of the response."[13]

Another report commissioned by Kan during the first week of the crisis and delivered two weeks later backed his fears of a worst-case scenario.[14] The 15-page report by the Japan Atomic Energy Commission warned that evacuation orders would have to be issued to residents living within 155 miles of the Daiichi plant if the situation there spiraled out of control, a radius that would have included the metropolitan Tokyo area. The warning was kept secret until 2012.

Testimony to the Diet by the three major political players in the disaster—Kan, Edano, and Kaieda—subsequently made it clear that they believed the utility was preparing to completely abandon the plant. All three told a Diet panel investigating the nuclear disaster that they could not recall any mention of leaving behind a skeleton crew. Recalling a telephone conversation with TEPCO president Masataka Shimizu on the early morning of March 15, Edano said he warned that the disaster would be "unstoppable" if TEPCO pulled out all of its workers. "Shimizu stammered, so it was clear that he did not intend to leave some workers (to contain the accident)," he said.[15]

The bizarre behavior of the utility's president—effectively abandoning his post, making himself inaccessible for much of March 2011, convincing the country's political leaders that he wanted a full pullout from the Daiichi plant—has left behind a deeply confusing legacy. The Diet report concluded that the "fundamental cause" of the crisis was a culture of "deference to and reliance on government authority." Somehow, Shimizu, TEPCO, and the regulators were afraid to relay the true state of things and said only what the government wanted to hear. Kan is seen by some in Japan as the nearest thing to a hero that the crisis produced precisely for very forcefully conveying his feelings and stepping into the leadership vacuum left by TEPCO and the regulators. But many, particularly in the political and nuclear establishment, consider him an arrogant meddler who misread the situation and made it worse. To some, his confrontational, aggressive approach was worse than unhelpful—it wasn't Japanese.

One hundred and fifty hours of internal video footage of the first five days of the crisis, reluctantly released by TEPCO in August 2012, failed to clear up what took place on March 14 and 15. The key scene of Kan's angry, arm-waving confrontation with TEPCO executives is silent, a result, said the utility, of "technical glitches." At times, managers are openly heard discussing evacuation. "At what time will all the workers be evacuating from the site?" a senior executive is heard asking then–executive vice president Sakae Muto at 7:55 p.m. on

March 14. "Tepco President Masataka Shimizu is heard saying at around 8:20 p.m. that 'a final evacuation has not been decided yet' and that he is in the process of checking with 'related authorities,' possibly referring to the office of then Prime Minister Naoto Kan."[16]

Somewhere between March and the hot summer months, Kan turned against nuclear power. It was not one single event that converted him, he later recalled, but the cumulative experience of dealing with Fukushima and the prospect of a world with double the number of reactors in the next two decades. "It was providence that the disaster was not worse," he said afterward, when he had time to consider what took place. "We were being pushed back by this unseen enemy and on the very brink of disaster on March 15–17. But finally we began to fight back. Tepco held firm, troops and fire fighters were sent in and we began to get water into the reactors."[17] Within months, a hostile conservative media would help hound him from office, accusing him of mishandling the crisis.

SIX

Telling the World

I beg you to help us.

—*Katsunobu Sakurai*

AS THE STRAIN OF THE NUCLEAR CRISIS BEGAN TO erode Japan's famous poise, reporting became increasingly controversial. Japan's big media more or less toed the official line, avoiding the word *meltdown,* for example, and initially shunning long-term critics of the nuclear industry, but some foreign outlets filed increasingly hyperbolic and speculative stories. One infamous report in the UK tabloid *Sun* compared Tokyo to a zombie "city of ghosts." Another by Fox News put a nuclear reactor in the heart of urban Shibuya.[1] Neither reporter had set foot in Japan.

Japan's foreign ministry led the criticism of "excessive" coverage, singling out the *Blade,* a local US newspaper from Toledo, Ohio, that ran a cartoon depicting three mushroom clouds, one each for Hiroshima, Nagasaki, and Fukushima.[2] *Newsweek Japan* took up the cudgel against shrill, alarmist *gaijin* reporters.[3] "The foreign media in Japan . . . has been put on a pedestal as the paragon of journalism, and was viewed as a source of credibility. The Great East Japan Earthquake shattered that myth," thundered editor Takashi Yokota. "The Western media failed to fulfill its mission during the disaster, hitting new lows with shoddy journalism as reporters were overtaken by the news and lost their composure."

Yokota accused foreign journalists of gross sensationalism after the first explosion at the Fukushima Daiichi nuclear plant, which quickly turned the story into "Japan's Chernobyl." This was the headline at *Newsweek* a week before the Japanese government officially raised Fukushima to the same level as the 1986 disaster. The *Wall Street Journal* also noted the "gulf" that Fukushima opened up in reporting, noting that while local journalists gave the sense that the "situation will be resolved," their foreign counterparts focused "on the other side—that this is getting out of control."[4]

In the week after the nuclear crisis erupted, Japan-based bloggers assembled a "wall of shame," citing dozens of foreign crimes against journalism.[5]

One problem with the foreign media was its lack of knowledge-able personnel in Japan. After the disaster, many journalists who had no knowledge of the country were dispatched to Japan. The resulting inaccurate or unbalanced reporting was criticized by local foreigners as well as Japanese. Jeff Kingston, director of Asian studies at Temple University's Japan Campus, was one of several critics who cited the "many egregious instances of . . . exaggeration and misrepresentation" fueled by what he called "parachute journalism."[6] For years, Japan's dreary, protracted economic decline had been a turnoff to distant editors, and the country had fallen off the media radar, eclipsed by fast-rising China. Tokyo-based hacks sometimes joked darkly that it would take a major disaster to revive Japan's newsworthiness. Disaster had duly arrived, and there were not enough reporters to cover it.

But hyperbolic reporting was not all imported. One of the most-cited examples was Japanese: *AERA* magazine's famous March 19 cover story showing a masked nuclear worker and the headline "Radiation Is Coming to Tokyo" was controversial enough to force an apology and the resignation of at least one columnist (though the headline was in fact correct).[7] Moreover, once the dust from the crisis settled, weekly Japanese magazines quickly sank their teeth into the nuclear industry and its administrators far more aggressively than the foreign media ever did.[8] Other magazines turned their critical gaze on the radiation issue, exposing government malfeasance and lies. *AERA* also revealed that local governments manipulated public opinion in support of reopening nuclear plants.

The Fukushima disaster revealed one of the major fault lines in Japanese journalism, which is between the mainstream newspapers and television companies and the less inhibited mass-selling weeklies and their ranks of freelancers. The subject was new, but the debate it amplified on the influence of the press club system had been going on for decades.[9] The system means that Japan's big newspapers and TV companies channel information directly from the nation's political, bureaucratic, and corporate elite to the media and the public beyond—in this case, from the government, TEPCO, and NISA. The system's critics say it locks Japan's most influential journalists into a symbiotic relationship with their sources and discourages them from investigation or independent lines of analysis and criticism.

Author and freelancer Takashi Uesugi was one of several who accused the local media of colluding with the government and the plant operator. "Tepco is a client of the media and the press clubs, being one of their biggest advertisers—so the press won't . . . say certain things," he said, citing their blackout of the meltdown that occurred in reactors one to three and the fact that the latter had a payload of lethal plutonium. Such statements were enough, he claimed, to get him banned from the Tokyo Broadcasting System (TBS) network in April 2011.[10] Former Washington TBS bureau chief Toyohiro Akiyama, who owned a farm in Fukushima, made a similar assessment, accusing the mass media of being a mouthpiece for the government and the power company.[11]

There is strong evidence for claims of structural bias. Before the Fukushima disaster, Japan's power-supply industry, collectively, was Japan's biggest advertiser, spending 88 billion yen (roughly US$1 billion) a year, according to the Nikkei Advertising Research Institute.[12] TEPCO's 24.4 billion yen alone is roughly half of what a global firm as large as Toyota spends in a year. Many supposedly neutral journalists were tied to the industry in complex ways: senior *Yomiuri* editorial and science writer, Masao Nakamura, for example, was an advisor to the Central Research Institute of Electric Power Industry; journalists from the *Nikkei* and *Mainichi* newspapers went on to work for pronuclear organizations and publications.[13] Before the Fukushima crisis began, TEPCO's largesse may have helped silence even the most liberal of potential critics.[14] That industry financial clout, combined for decades with the press club system, surely helped discourage investigative reporting and keep concerns about nuclear power and critics of dangerous plants like Hamaoka and Fukushima well below the media radar.[15]

In Fukushima itself, however, at least until the government made it illegal in late April 2011 to enter the 12-mile irradiated evacuee zone, access was almost unlimited. Locals there told immensely poignant stories, expressing bewilderment and anger at their fate at the hands of a plant that didn't deliver a single watt of electricity to Fukushima. In an echo of Bhopal, Chernobyl, and other accidents steeped in epic corporate hubris, they felt that they had been manipulated, lied to, and finally abandoned by TEPCO. Still, some were determined to stay rather than abandon houses and farms that had been in their families, in some cases since the Meiji era (1868–1912) or before.

In Minamisoma, Mayor Sakurai watched in despair through March as the crisis spun out of control. By the end of the month, there were fewer than 20,000 people left in the city out of an original population of 71,000. The earthquake and tsunami killed or left missing over 900; the remainder fled from the threat of radiation. Journalists from the big Japanese media pulled back to areas considered safe from the (then unconfirmed) radiation fallout. They returned some 40 days later.[16] The decision, he says, significantly worsened the situation for the city. "We were abandoned, so there was no way to tell the country or the world what was happening."

Once the explosions began at the Daiichi plant, regular deliveries of food and fuel began to dwindle. Information about the state of the power plant was gleaned from the television, which relied on openly pronuclear experts to explain what was happening to the six reactors. The most prominent and heavily rotated was Naoto Sekimura, a vice dean of the Graduate School of Engineering at the University of Tokyo and a consultant with METI's (Ministry of Economy, Trade, and Industry) Advisory Committee for Natural Resources and Energy. Sekimura previously wrote reports verifying the structural soundness of the Fukushima plant and had signed off on a ten-year extension for the number one reactor.[17] The comments of other pronuclear scientists were also heavily reported—notably those of Madarame Haruki, the chairman of the Nuclear Safety Commission of Japan—to the exclusion of alternative voices.

Most of Sekimura's on-air comments reflected his close ties to the industry and were, he admitted later, drawn from his contacts inside TEPCO. "Residents near the power station should stay calm," he said on March 12, shortly before the first hydrogen explosion. "Most of the fuel remains inside the reactor, which has stopped operation and is being cooled." It would take TEPCO two months to admit that the uranium fuel inside the number one reactor had by this time already completely melted. Sekimura assured his audience that a major radioactive disaster was "unlikely." A short time later, the explosion destroyed the concrete building housing reactor one, contaminating the surrounding countryside and sea.

Japan's public service broadcaster NHK has a network of 54 bureaus throughout Japan, thousands of journalists, 14 helicopters, and over 60 mobile broadcasting units. It reaches 50 million households and is among the most trusted sources in the world. Experts have compared its clout to ABC, NBC, and CBS News combined.[18] With that network, and its exclusive access to disaster information,

NHK did a superb job of relaying information from government and corporate sources but did less well in analyzing it. Long-term critics of the nuclear industry, such as Ikuro Anzai, a radiation specialist and former professor at the University of Tokyo's nuclear engineering department, or Kyoto University researcher Hiroaki Koide, were largely ignored. In a postmortem of NHK's March/April coverage, Noriyuki Ogi, head of broadcasting during the disaster, said this of the nuclear crisis: "Overwhelmingly the problem was lack of information. Even TEPCO and the government didn't know the whole picture. We didn't have enough time to evaluate their reports and so we didn't know how far we should go in telling the dangers of the situation. We were relying on TEPCO and the government and because they were not sure, we were not sure."[19]

Ogi said that NHK had gone above and beyond the call of duty: "On the afternoon of March 12, the police only reported that the sound of an explosion had been heard. Tepco, NISA and the government said nothing. Looking at the screen, our reporter noticed what was happening and said, 'Just in case, anyone who is outside please go inside and stay out of the rain.' Even though we didn't have any proof, we went further than we needed to."[20]

Mayor Sakurai fretted about those who couldn't flee. His remaining constituents included many elderly and bedridden people who faced starvation because delivery trucks were staying away from the city. Skeleton staff stayed on at the hospitals with their patients, refusing to leave. Television reporters occasionally called from Fukushima City or Tokyo for updates, but with so many other stories clamoring for attention, there seemed no way to impress on them how desperate the situation was. There would be no direct word from TEPCO on the state of the Daiichi plant for 22 days. "We're isolated. The government doesn't tell us anything," he told the BBC during the first week after the crisis began. "They're leaving us to die." He knew that many of his own staff also wanted to flee with their families. So on March 24, the mayor sat in front of a camcorder in his office and recorded an 11-minute video that was uploaded to YouTube with English subtitles. "We are not getting enough information from the government and Tokyo Electric Power Co.," said the exhausted mayor, pleading for volunteers to come and help "at their own risk." Journalists must come and see the situation for themselves and stop relying on telephone interviews. "All the convenience stores and supermarkets where people buy everyday goods are shut down. Citizens are almost being driven into starvation. . . . The people are literally drying up as if they are under starvation tactics. . . . I beg you to help us."[21]

The video registered more than 200,000 hits in the following week and attracted tons of aid from the United States and around the world. It also drew a stream of freelance Japanese and foreign reporters who made Sakurai an emblematic figure of the challenge to blundering and incompetent officialdom during the disaster. Sakurai felt that journalists should have gotten the message out instead of leaving to protect themselves. "I appreciate that there were dangers, but we had many people who stayed behind, and in my view the journalists should have stayed, too." What struck him about the Minamisoma episode is how the Japanese journalists acted together, like a retreating army. Speaking anonymously, a reporter for one of the major newspapers said that he and his colleagues were left with no choice once they were told to leave. "There was some discussion but in the end we agreed that it would be safer to report from Fukushima city." There was no conscious collective decision. It happened almost by osmosis. When they returned, he added, Mayor Sakurai had berated them. "He said the foreign media and freelancers came in droves to report what happened. What about you?"

The reporting of the Minamisoma story demonstrated the discipline of the Japanese press corps. Satoru Masuyama, a director with NHK's science and culture division, called the decision to pull out of Minamisoma a case of individual versus corporate responsibility. Reporters at a Japanese company will not take risks by themselves; they will wait for instructions. And the company will not send its workers off without proper preparation or protective gear. Many critics would later question why none of the big media broke ranks in the interests of their readers. "I subscribe to four major national newspapers, but I cannot tell which newspaper I am reading in relation to articles about the nuclear accident," Tatsuru Uchida, a professor at Kobe College, told the *Asahi* newspaper.[22] Uchida said there was no attempt to bring out a unique angle because of the fear of reporting something different from the other papers. For him, it resembled what happened during World War II, when the media repeatedly lied about the nation's disastrous military campaign.

After Sakurai's video, life inside the 12-mile zone around the power plant became one of the most sought-after stories in the world. The government had steadily strengthened this zone from advising evacuation on March 11 to ordering evacuation for 70,000 to 80,000 people later that week, while another 136,000 people in the zone 12 to nearly 20 miles away were told to stay in their homes. The government directive was widely criticized by Fukushima residents and some sections of the media as arbitrary and unscientific. Eventually, several villages outside the zone would also be evacuated, such as Iitate. A small number of mainly elderly people stayed behind, refusing to leave homes that had often been in their families for generations. Not surprisingly, there was enormous global interest in their story and its disturbing echoes of the Chernobyl catastrophe 25 years earlier.

In late March, a trickle of foreign journalists braved radiation inside the zone. *Newsweek*'s Joshua Hammer described it as the "*Twilight Zone* crossed with *The Day After*—an apocalyptic vision of life in the nuclear age." Daniel Howden, from David McNeill's newspaper, the *Independent,* drove right to the gates of the plant, encountering deserted homes, stray pets, and nervous nuclear workers along the way. But he was unable to find interviewees inside the zone, so a few days later, McNeill followed him and talked to several holdouts. Neither of them encountered a single Japanese reporter inside the exclusion zone, despite the fact that it was not yet illegal to be there. Some would begin reporting from the area much later, after receiving government clearance—the *Asahi* sent its first dispatch on April 25 when its reporters accompanied the commissioner general of the National Police Agency. Later, they would explain why they stayed away and—with the exception of approved government excursions—continued to stay away. "Journalists are employees and their companies have to protect them from dangers," explained Keiichi Satō, a deputy editor with the news division of Nippon TV. "Reporters like myself might want to go into that zone and get the story, and there was internal debate about that, but there isn't much personal freedom inside big media companies. We were told by our superiors that it was dangerous, so going in by ourselves would mean breaking that rule. It would mean nothing less than quitting the company."[23]

Outside Japan, foreign correspondents are increasingly retained by newspapers on casual contracts or as stringers, reflecting both shrinking budgets and the declining importance of all but a handful of must-have global stories. Reporters like Hammer and Howden, brought over from their normal beats (in the Middle East and Africa, respectively) precisely for their skills and bravery in difficult assignments, are under a lot of unspoken pressure to justify the expense of getting them there. They are expected to use their skills of interpretation and analysis in situations where they don't always know what is going on. In addition, their stories are bylined, bringing a certain amount of individual glory in the event of a scoop. That background, the reporters' lack of specific knowledge about nuclear power and their unfamiliarity with Japan, helps explain the occasional sensationalist dispatch.

In contrast, reporters for Japan's big media are generally staffers, usually embedded in organizations with a strict line of command and lifetime employment. The emphasis at these companies is on a descriptive, fact-based style relying on official sources. Investigative reporting is limited, and the individual reputation of each reporter is considered less important than those of their Western counterparts. Most of the stories carried in the Japanese newspapers are not bylined. In practice, this means that the best investigative reporting in Japan is often done by freelancers.[24]

It is not difficult from this context to see two very different dynamics at work. Unlike their foreign counterparts, Japanese reporters for the big media had little to gain from breaking ranks and disregarding government warnings on the dangers of reporting close to the nuclear plant. Moreover, the cartel-like behavior of the Japanese companies meant they did not have to fear being trumped by rivals.[25] In particularly dangerous situations, managers of TV networks and newspapers will form agreements (known as *hōdō kyōtei*) in effect to collectively keep their reporters out of harm's way. Teddy Jimbo, founder of the pioneering Internet broadcaster Video News Network, explains, "Once the five or six big firms come to an agreement with their competitors not to do anything, they don't have to be worried about being scooped or challenged."[26] Jimbo says that the eruption of Mount Unzen in 1991 and the 2003 invasion of Iraq, both of which led to fatalities among Japanese journalists, copper-fastened these agreements—one reason that so few Japanese reporters can be seen in recent conflict zones such as Burma, Thailand, or Afghanistan. The *Times*' Asia bureau chief, Richard Lloyd Parry, who has reported from all of those conflicts, sums up his observations thus: "Japanese journalists are among the most risk-averse in the world."[27]

Frustrated by the lack of information from around the plant, in the end, Jimbo took his camera and dosimeters into the 12-mile zone on April 2 and, like Sakurai, uploaded a report on YouTube that scored almost a million hits. He was the first Japanese reporter to bring television images from Futaba and other abandoned towns. "For freelance journalists, it's not hard to beat the big companies because you quickly learn where their line is," he said. "As a journalist I needed to go in and find out what was happening. Any real journalist would want to do that."[28] He later sold some of his footage to three of the big Japanese TV networks: NHK, NTV, and TBS. "For two months they were showing graphics on TV about what was happening," he said. "All they did was quote experts, TEPCO and others from the 'nuclear village.' So that meant that everything they showed was wrong."[29]

View of Rikuzentakata, a coastline town destroyed by the tsunami. Few structures in the center of town near the sea withstood the tsunami. In the foreground is a gateway to Jodo-ji Buddhist temple. Photographer: Lucy Birmingham

A solitary black pine stands behind a destroyed floodgate in Rikuzentakata, Iwate Prefecture. The pine tree was the only one left standing out of nearly 70,000 trees after the tsunami destroyed the town. The tree quickly became a symbol of hope for the region but has since succumbed to the salty soil, despite the efforts of volunteers and city officials. Photographer: Robert Gilhooly

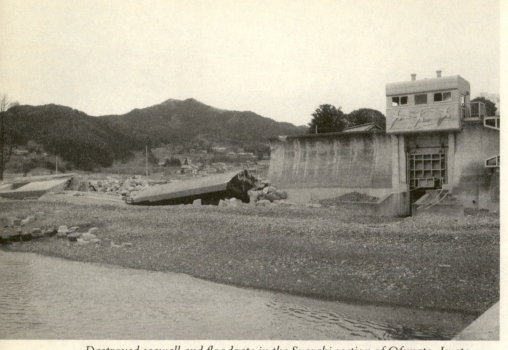

Destroyed seawall and floodgate in the Suezaki section of Ofunato, Iwate Prefecture. Photographer: Lucy Birmingham

Around 300 people take shelter at a school gymnasium in Iwaki, Fukushima Prefecture, on March 12, 2011, about 21 miles from the Fukushima Daiichi Nuclear Power Plant accident. Photographer: Robert Gilhooly

A clock lies among the debris in Otsuchi Prefecture, showing the time the tsunami hit the town. Photographer: Robert Gilhooly

A few days after the quake, a woman looks through a list of the names of people who have been delivered to a temporary morgue inside the Ishinomaki Municipal Gymnasium in Ishinomaki, Miyagi Prefecture. Photographer: Robert Gilhooly

A sign indicating the end of a tsunami inundation area lies among the debris after the megatsunami in Minamisanriku, Miyagi Prefecture. Photographer: Robert Gilhooly

A pleasure boat sits atop a building after being washed inland by the tsunami in Otsuchi, Iwate Prefecture. Photographer: Robert Gilhooly

Mourners walk through the rubble of their former homes in Rikuzentakata, Iwate Prefecture. Photographer: Robert Gilhooly

A man cycles past the Asia Symphony cargo ship that was swept ashore by the tsunami in the port of Kamaishi, Iwate Prefecture. Photographer: Robert Gilhooly

Anti-nuclear demonstration, Tokyo, April 10, 2011. Photographer: David McNeill

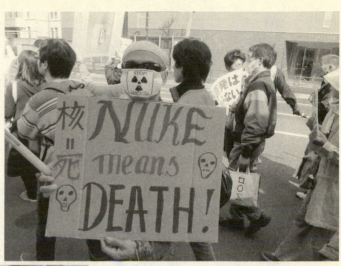

Masked security guard at the Fukushima Daiichi Nuclear Power Plant angrily waving away coauthor David McNeill, days after the quake. Photographer: David McNeill

The cracked wall of the onsite visitor's center at the Fukushima Daiichi Nuclear Power Plant. Photographer: David McNeill

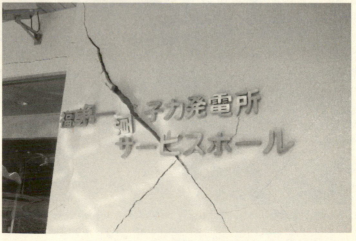

English teacher David Chumreonlert inside his elementary school gymnasium where he saved students, parents, and elderly during the tsunami flooding. The clock shows the time it stopped just after the earthquake hit. Higashi-Matsushima, Miyagi Prefecture. Photographer: Hiroshi Sato

Setsuko Uwabe stands next to the entrance of the nursery school where she worked as a cook before the tsunami destroyed it. Photographer: Hiroshi Sato

Fisherman Yoshio Ichida looks out across the ocean where he fished before the tsunami and radioactivity from the Fukushima Daiichi Nuclear Power Plant contaminated the waters. Photographer: Hiroshi Sato

Toru Saito standing on the empty land where his house stood before it was destroyed by the tsunami. Photographer: Hiroshi Sato

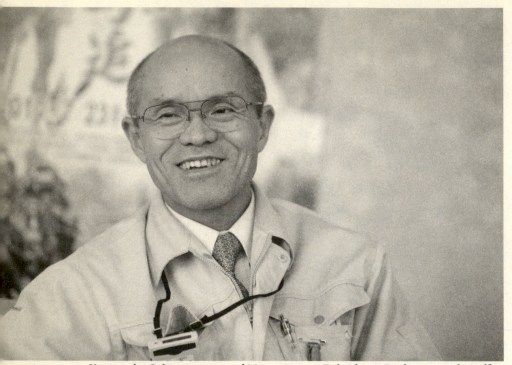

Katsunobu Sakurai, mayor of Minamisoma, Fukushima Prefecture, in his office a few months after the quake. Photographer: Hiroshi Sato

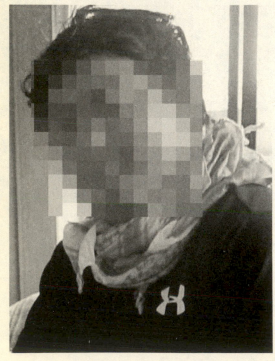

TEPCO worker Kai Watanabe at the Fukushima Daiichi Nuclear Power Plant, Okuma, Fukushima Prefecture. Photographer: David McNeill

SEVEN

Flyjin

The State Department strongly urges U.S. citizens to defer travel to Japan at this time and those in Japan should consider departing.

—US Embassy

THROUGHOUT THE TWO DAYS THAT HE WAS STRANDED at Nobiru Elementary School, one of the thoughts weighing heaviest on David Chumreonlert's mind was how to let his family and friends back in the States know that he was okay. Electricity and communication lines were cut off throughout the area. He knew they would be worried sick and fearing the worst. There was no way for them to know if he was alive or dead. The brief e-mail he had sent to his friend just seconds after the quake had hit was his last communication with the outside world.

Thanks to the perseverance of the local board of education, he was able to get back to his apartment nearby in Higashi-Matsushima on Sunday, March 13. Most of the roadways were still flooded and clogged with debris. But a staff member had managed to navigate the back roads by car to get him at the elementary school. The tsunami waves had stopped just across the street from his building. He couldn't believe his luck. Inside his apartment, the quake had catapulted almost everything across the floor, making a huge mess, but otherwise the rooms were intact.

To contact his family, David knew he needed to get to Sendai about 30 miles away. The city was still mostly shut down, but electricity had returned to certain areas. He would then be able to get telephone and Internet access. But how was he to get there? He was finally able to hitch a ride to the city that Wednesday with the CBS *60 Minutes* crew that had interviewed him that day for their disaster coverage.[1] He first headed to his church, but no luck. Electricity there was still down. He then hurried to Greg Lekich's apartment. He had heard that Greg, a friend and fellow English teacher, was opening his home to those who needed Internet access and a place to stay.

Packed with about 15 foreigners, Greg's apartment looked like a miniconvention for international refugees. Everyone was online connecting with friends, family, and the many worried people who were trying to contact them. David's cell phone was ruined in the tsunami water, so he immediately Skyped his brother's landline phone in Houston. It was the middle of the night when his brother answered the call.

"Hey, it's David. I'm okay. I'm alive!" David said, trying to quickly reassure him.

"What?" answered his groggy brother. He had not recognized the number that flashed on the phone.

"It's me, David. I'm alive!"

"Oh!" he answered, slowly emerging from a deep sleep. "Is everything all right? Are you safe?" He was so relieved to hear David's voice.

"Just tell everyone that I'm okay. That I'm alive." His brother agreed to contact the rest of the family but did not have their mother's number in Bangkok where she was visiting relatives. David turned to Facebook and immediately posted a message asking if anyone had her contact information. A worried family friend who had been checking David's page every day happened to be online and replied back within about 20 minutes with his mother's number.

David learned later that many friends had been checking his Facebook page daily for news and posting any information they could get. One friend was so worried that he checked David's page throughout the day from the moment he woke up. Someone had also listed him on Google Person Finder. But Higashi-Matsushima was so isolated that no one could get news of him. Many feared he had died in the tsunami.

His parents and relatives had quickly contacted both the US and Thai embassies to request a search. David thinks it was about a week after the quake and tsunami that US Embassy staff came knocking at his apartment door. They checked his name off their list, gave him a number to call if he needed any help, and assured him that his family would be contacted. Fortunately by that time, he had already been in touch with his family.

"Are you all right?" his mother cried when she heard David's voice. "Aren't you worried about the radiation? When are you coming home?" He tried his best to calm her fears, but she was beside herself. She wanted him to get out of Japan. After watching and reading the news, it seemed she knew more about the Fukushima plant accident and radiation plumes than he did. Without news access, he had not realized the seriousness of the situation. He had been busy just trying to survive with limited food and water.

David explained that Higashi-Matsushima and Sendai were about 62 miles from the Fukushima accident and far enough away not to be affected by the radiation, but she wasn't convinced. "I can send you a ticket right away. Please come back home!" she pleaded. He knew his mother too well. This wasn't a request. It was a demand. She wasn't going to change her mind easily on this one. As everyone in the family agreed, she was the boss. "I'd like to stay," he explained calmly. "There are so many people who need help right now."

She understood his feelings. It was part of their church's calling to help those in need. But this was different. This was her son, in the same region where experts were saying a nuclear meltdown had occurred. David tried to reassure her. He promised he would come back to the States in a few weeks.

David returned to Higashi-Matsushima from Sendai for about a week and stayed with a Japanese family. The husband was with the

Japan Self-Defense Forces and out all day, busy with rescue efforts. David was able to help procure water and food for the family. Potable water was sourced from a truck driving through the neighborhoods at unpredictable times. Whenever word got out that the truck was nearby, they would quickly gather all the empty plastic bottles they could find and run to wait in the long line to fill up. At the MaxValu supermarket nearby, David would line up by 6:00 A.M., along with about a hundred others, to purchase the six allotted items. Without refrigeration, there was virtually no fresh food. Cash registers were inoperable. The staff had to calculate the costs by hand.

It was a time-consuming, frustrating experience, but there were few audible grumbles among the crowds. David was amazed at the level of organization and cooperation. The Japanese seemed to work together, suppressing ego and selfish desires at precisely the time when they might be expected to flare. Many commentators have noted how group harmony, often under duress, is highly valued in Japan. Now it was faced with a crisis unlike any other. The Japanese had survived horrific disasters before. Surely their ingrained sense of self-sacrifice and unified purpose would enable them to overcome the calamities at hand.

Unity, of course, was key. Or was it? There was deep uncertainty about the consequences of the nuclear plant explosions and radiation exposure. A creeping sense of conspiracy and cover-up was unfolding as the government, plant operator TEPCO, and so-called experts contradicted themselves and international sources. People were getting angry. The stress was unbearable. In evacuation centers tempers flared and fighting occurred.

For many foreign residents like David, unity was not an option. With worried family back home begging them to get out of Japan and with radiation effects unknown, it made sense to leave. It was also school break for Japanese and most international schools. David's teaching contract had already been renewed, so he knew he would be back in April at the start of the new school year.

Getting out, though, turned into an unexpected trial. Without a car or local transportation, David's only option was a treacherous four-hour bicycle ride from his home back to Sendai, traversing cracked highways and debris-filled roadways. From there he caught a Shinkansen bullet train packed with women and children heading to southwestern regions. Since the school year had finished, they could have been on vacation, but they were also likely heading away from radiation concerns in Tokyo. The capital is only 155 miles away from the Fukushima plant. Radiation fallout, it seemed, depended upon the unpredictable strength and direction of the wind.

David's first destination was Kobe, where he would go to a friend's wedding, then Osaka International Airport nearby to get back to the States. His flight out on March 30 cemented his spot among the many newly dubbed *"flyjin."* A combination of *fly* and the word *gaijin* (literally, "outsider"), it first surfaced from the foreign community and became a convenient way to deride foreigners who appeared to be running away from their jobs and responsibilities. A combination of sympathy and anger emerged from both foreigners and Japanese left to cover the extra workload. For *flyjin*, the choice to leave was multilayered and made sense.

THE SHOULD-I-RUN-OR-STAY DEBATE began to surface on Sunday morning, March 13, when the Japan Meteorological Agency announced that there was a 70 percent chance that another major earthquake with a magnitude of 7 could hit the Tokyo metropolitan area within three days.

Multinational companies with offices in Tokyo, many concerned about insurance ramifications, were recommending their foreign employees to leave. One German executive with an international consulting firm in Tokyo said he would be leaving for Hong Kong with his wife and three young children on Monday, the fourteenth. "I am more concerned about the reactor situation than earthquakes, but the combination of the two and aftershocks is decreasing our confidence level," he admitted. "My wife is having a very hard time with the strong aftershocks. She really wants to leave."

His company had quickly done its homework on the risks involved. They had consulted with Japanese nuclear physicists who determined that the nuclear plant accident was a level 4 out of 7, so there was no immediate risk. Later, the assessment proved wrong after Fukushima was given the maximum level 7 rating, equal to the 1986 Chernobyl nuclear accident. Some companies tried to stop the exodus with threats. One high-powered British senior executive with a worldwide communications consultancy, stated flatly, "If any foreigner in our Tokyo office leaves because of this disaster, they're fired."

But mass fear is difficult to control. By the afternoon of Sunday, the thirteenth, rumors were spreading that European embassies were quietly contacting their citizens with young children and recommending that they prepare to leave. It may have been a responsible and natural response to protect their citizens, but it would have awkward diplomatic ramifications and turn into a logistical nightmare.

When contacted early Sunday afternoon, one European Union ambassador lambasted the rumormongers. "Embassies recommending they leave Japan? That's absolutely not true. It would be saying we don't trust the Japanese government," he said confidently. "What is true is that on our embassy websites there is travel advice. All of our embassies more or less have been telling their people not to come to Japan if they don't have essential business at this time. But there has been some panic here." By early evening, however, the tide had turned.

The French and German Embassies posted an online warning to their thousands of citizens in Japan to leave Tokyo and the surrounding Kanto region. The European ambassador quickly rescinded his earlier comments. "We'll be following the French lead," he clarified.

The French were quickly provided with flights out and were among the first to leave. Ironically, a few weeks later on March 30, French president Nicolas Sarkozy came to Japan to pledge faith in nuclear power, with executives from the country's powerful state-owned nuclear company AREVA hovering close behind. Standing beside Sarkozy, Japan's exhausted-looking prime minister nodded his head, even as he was becoming an antinuclear convert. Three months later, Kan would call for a nuclear-free Japan.

Soon after the earthquake and tsunami on March 11, the US Embassy sent out travel alerts to its citizens in Japan, with an update on the thirteenth. Along with emergency information, the recommendation was to avoid travel to Japan. Also included was the recommendation by NISA that people "who live within 20 kilometers [12 miles] of the Fukushima Nuclear Power Plant in Okumacho evacuate the area immediately." The memo went on to say, "Japanese authorities have confirmed that the situation remains serious. U.S. citizens residing or traveling in Fukushima Prefecture should follow NISA instructions to evacuate and comply with Japanese government personnel on the ground."[2]

By Wednesday, March 16, the warnings to US citizens in Japan had become far starker. "The State Department strongly urges U.S. citizens to defer travel to Japan at this time and those in Japan should consider departing," wrote the US Embassy, which arranged one-way flights out of the country to the "safe haven" destinations of Taipei, Taiwan and Seoul, South Korea. Passengers were required to sign an agreement to pay $3,000 per person per flight, an amount determined by the airline. Few citizens, however, took advantage of the service. The embassy also arranged chartered buses for citizens in Tohoku wanting to leave the region at a cost of $50.00 per person.[3]

Families of government personnel in Tokyo and nearby Nagoya and Yokohama were also given the option to leave with the standard voluntary departure period of 30 days. In contrast, the ranks of US Embassy staff swelled as about 150 government experts rushed to Tokyo to deal with the Fukushima crisis.

Japan's beleaguered government began to contemplate the unthinkable—evacuating Tokyo. It was late March, two weeks after the quake, and Prime Minister Kan put together a worst-case scenario in which he was told to consider evacuating everyone residing within 125 to 155 miles of the plant. This included Tokyo and adjacent areas across the island of Honshu to the Sea of Japan. The recommendation was not made public.

For Americans, evacuations became a reality. A March 16 travel warning from the US Embassy upped the ante by recommending that US citizens living within 50 miles of the Daiichi nuclear plant evacuate or stay indoors. The Japanese government had determined that 12 miles was sufficient and had evacuated all residents within it. The US recommendation was five times larger than the zone established for emergencies at US nuclear power plants. Did the US government know something the Japanese did not?

Ambassador John Roos later explained the comparative safety concerns facing both Japan and the United States. Experts at the Nuclear Regulatory Commission (NRC) going through the analyses were asking what the United States would have done if confronted with a similar situation.[4]

The NRC, it turns out, was depending heavily on unreliable TEPCO news releases, the International Atomic Energy Agency, and the news media. Transcripts of telephone conversations released by the NRC nearly a year later revealed the confusion that occurred.

During those first few days of the unfolding crisis, the agency struggled to gain accurate information.

In Washington, DC, in the early morning hours of March 11, NRC chairman Gregory Jaczko and his staff were discussing potential damage from the tsunami on US nuclear power plants. The Diablo Canyon plant in California was a real concern. It was midmorning when they received word from the International Atomic Energy Agency that the Fukushima Daiichi plant's backup power and cooling systems had failed. That afternoon, NRC official Bill Ruland said during a conference call that the plant complex was suffering a blackout. "We're about at the time where they could start to see core damage," he said, adding that information was "meager." Later, during a conference call with Jaczko, he said, "We're dying in a sea of silence here, actually."[5]

Jaczko's controversial decision to set the evacuation radius at 50 miles was based in part on what they understood was an assessment by Japanese officials. Jaczko told Congress that the plant's reactor four spent-fuel pool was dry and that this would increase the chances of a catastrophic radioactive release. Later, an NRC official told Jaczko that Japanese officials who had initially indicated that the pool was dry had changed their minds.

Jaczko's top-ranking aide in Japan, Charles Casto, expressed frustration that information provided by the Japanese was limited. At one point, they were relying on a short video of the plant made by helicopter showing steam coming out of the building. In the United States there would be a direct line of communication with the reactor's operator and a clearer picture of what was happening, said an NRC deputy director.

The transcripts on March 16 showed that NRC staff members debated the evacuation issue and how it would be carried out. That day, Bill Borchardt, the NRC's executive director for operations, told Jaczko, "If this happened in the U.S., we would go out to 50 miles."

The recorded conversations also showed that US officials quickly understood the depth of the crisis. Jaczko correctly predicted the outcome. "At this point I would see a worst scenario probably being three reactors eventually having, for lack of a better term, a meltdown," he told White House officials.[6]

On Thursday, March 17, the US military authorized evacuations of eligible family members. Military commanders stationed in Japan were telling families that the Fukushima accident did not pose a significant radiation threat. But about 10,000 took the free flights. The Department of Defense would pay nearly $35 million to help American military families flee. The navy paid the most with about $14.4 million in reimbursements because there are two navy bases within about 150 miles of the Fukushima plant. Some families who headed to the United States stayed beyond the one-month evacuation period to allow children to finish the school year. Some chose Asian beach resort getaways.

Statements from the embassies and a barrage of sensationalized media reporting kicked the fear factor and exodus into high gear. The question on everyone's mind: Will radiation from the meltdown reach the capital? Statements on radiation leakage amounts and acceptable limits were contradicting expert opinions within Japan and abroad.

Later on March 23, Japanese authorities warned that Tokyo tap water was unsafe for infants after radioactive iodine was discovered in the city's water supply. According to Japanese news media reports that day, chief cabinet secretary Yukio Edano said that the radiation was being carried on the air from the Fukushima plant. "Because it's raining, it's possible that a lot of places will be affected," he said. But he added that consuming the water a few times would not cause long-term effects.[7] The Health Ministry also offered a confusing response, saying that the risk for infants was unlikely but that tap water should be avoided and not used for infant formula. Prime Minister Kan warned earlier in the day that the public should avoid farm produce from areas near the Fukushima plant. These statements were not only a confirmation that radiation was contaminating water and food, but also the first nail in the coffin for farmers and fishermen in the Tohoku region.

As the Fukushima crisis unfolded that first week, Tokyo's immigration office bulged with frantic activity as foreigners scrambled to get out of Japan. Applications for reentry permits were surging, indicating that many planned to return. Lines were forming at long-distance and bullet train ticket counters. Flights out of Narita International Airport were booked up fast, forcing people to quickly find alternatives. Corporations arranged chartered fights out for their executives and families. Indelible photos of Chinese, donning protective face masks and lining up obediently at airport ticket counters, were a clear message that this was not just an American- and European-centric phenomenon.

The domestic refuge of choice became the Kansai region, with its business hubs of Osaka and Kobe, about 248 miles southwest of Tokyo. Hotels that were empty after Japanese cancelled their reservations quickly filled with foreign guests. Hotel suites were booked up by companies setting up temporary operations. One journalist commented with a laugh that the lobby of the Hyatt Regency Hotel in Kyoto had turned into a bad Sunday at National Azabu Supermarket, a favorite among Tokyo expats. Southeast Asian vacation getaways were among the favorite escape venues. Hong Kong became the business hub alternative.

The *flyjin* mass exodus out of Japan was not an illusion, and yet the actual numbers show that it applied to a small percentage of foreigners. According to a Ministry of Justice announcement on April 15, 2011, the number of foreigners leaving Japan that first month during the crisis came to 531,000. Of those who left, 302,000 had obtained reentry permits, meaning they planned to return.

FOR MANY FOREIGN RESIDENTS, leaving Tokyo was not a deliberate escape. The timing was strangely convenient. The disasters occurred about one week before the international schools had spring holiday. "Families typically scatter in Asia for spring break," says Karen Thomas, an American, long-term resident. "Everyone already had their tickets for Bali or Thailand. They just accelerated what they were going to do anyway." For Karen, her husband Jack Bird, and their four sons, the original plan was to spend spring break in Tokyo. They considered it their home. Two of their sons were born there. The younger two were at the American School in Japan (ASIJ). The older two, both graduates of ASIJ, were attending colleges in California.

Karen and Jack first came to Tokyo in the early 1990s on an expat package with Jack's company. They have stayed pretty much ever since. The family is among an unusual breed, with roots now firmly planted in both the United States and Japan.

Immediately after the quake on Friday, ASIJ began contacting parents. The temblor happened at the end of the school day, and many students had train and bus rides to complete. On Saturday, families were informed of the board of directors' decision to suspend school on Monday for one day. On Sunday, however, all parents were contacted again. Based on the many unknowns, including the unfolding nuclear crisis, ASIJ decided to suspend classes for the remaining four days before spring break. School would be reconvened on Monday, the twenty-eighth, after the scheduled break. They assured everyone that communication with families would be open throughout the two weeks.

Karen spent the next few days organizing emergency goods and food for her family and the affected areas. There was a big push to collect supplies, such as canned foods, rice, toothpaste, underwear, adult diapers, socks, and baby formula. Karen took bags of nonperishables to the Tokyo American Club and Second Harvest, a food bank and distribution nonprofit organization (NPO) that was accepting donations. Allied Pickfords, an international mover, and the ASIJ school buses were ferrying supplies to the disaster areas from various distribution points. Volunteers were working to get to the disaster area and ensure that those in need would get the items.

By Wednesday, with aftershocks continuing and frightening media reports escalating, Karen began to reconsider their decision to stay. "It was unnerving to watch the news on the big networks like CNN and BBC," she says. "If there was a tremor while the reporters were on camera, you experienced their panic. They also showed the tsunami footage over and over again and the scale was simply unfathomable." The contradictions among so-called experts were especially worrisome. She did not know whom to believe. "We were hearing, 'The plumes of smoke coming from the plant are safe,' and 'The plumes are radioactive and heading towards Tokyo!'" Additionally, local stores were empty of water, batteries, candles, and all prepared foods. Many gas stations had run out of petrol. Karen says she even heard reports that some airlines were refusing to fly into the airports because of radiation concerns.

Karen finally had to shut off the news after a few days. She just could not cope. "Shutting it off wasn't a disregard for Japan, or Katrina, 9/11, or any disaster," she says. "It's a protective mechanism you have to put in place to function as a human being."

It was Thursday, March 17, when Karen decided she had had enough of the emotional strain and physical danger. The intense aftershocks were becoming unbearable. Their sixth-floor apartment felt like it was in constant motion. She was used to earthquakes, having lived in Tokyo for so long. But the aftershock that threw her out of bed at 4:00 or 5:00 that morning tipped the scales. She and Jack guessed it was a magnitude 6 or 7. If the epicenter was in Tokyo, it could have been catastrophic. They left on a midnight flight that night for California. At that point, Karen's nerves were frayed. "We had the option to leave and had no idea how long that option would remain available to us," she says. "It was a heart-wrenching decision to leave our friends and the country that was our home not knowing, based on the circumstances, whether we would ever return. It might have been like being a foreigner in New York City during 9/11."

During the two weeks that Karen and her family were gone, there was a flurry of e-mails flying back and forth between ASIJ parents and friends strategizing ways to help. By the second week, organizations were e-mailing Karen lists of things to buy and donate. When she returned to Tokyo on March 28, she brought back several suitcases with items that could be used immediately. The Tokyo American Club Women's Group, Hands on Tokyo, and ASIJ organized drives to collect donated items. She is still not sure if it would have been better to donate money. "Either way, there was a huge outpouring of kindness and generosity from everyone I knew, Japanese and foreign," she says. "Everyone opened up their hearts to help."

For cash-strapped foreigners stranded in the hard-hit regions, getting out and back to their country was a challenge. Groups like the Geneva-based International Organization for Migration pitched in to help. By late March their liaison mission in Tokyo had assisted more than 100 foreigners to leave and was expected to help several thousand more. It was the first time for the refugee assistance organization to start a program in Japan.[8]

SOME FOREIGNERS LIVING IN THE HARD-HIT AREAS decided to stick it out and help. David Chumreonlert's friend Greg Lekich was teaching at a high school in Sendai on March 11 when the earthquake hit. Originally from Philadelphia, he was an ALT and had been teaching in the region's public schools since 2007.

Although electricity had gone out at Greg's high school after the jolting quake, they were able to get updates from a battery-operated radio. They knew a tsunami was coming. Fortunately the school was not located near the coast, so it was out of range. But it took Greg three hours to walk back to his apartment, closer to central Sendai, and the tsunami-hit area. "Nobody had any idea how terrible this was going to be," says Greg.

On the morning of March 12, there was still no power or water when friends came to his apartment by bicycle. With little information available they were curious to learn what was going on, so they headed for city hall where a TV was set up and watched the shocking news on NHK's national news program. Only then did they realize this was going to be a worldwide story. He was able to get on the Internet that night after power returned to his apartment. "On Facebook I had a million posts and realized everyone was worried," he says. "People didn't understand where exactly I was located."

With the help of friends staying at his apartment those first few weeks, Greg was able to split the daily survival tasks. Some waited in distribution lines for food and water. Some got information on certain people and sent word back to their families. Twenty-four foreign teachers were still unaccounted for, including David Chumreonlert. Friends who had mountain bicycles rode northward near the coast to the tsunami-hit towns of Ishinomaki, Matsushima, and Higashi-Matsushima where David lived. The friend who found David was able to let Greg know he was okay by e-mailing via the 3G network on his iPhone. "There was no other way we could have found him," says Greg. "Families abroad felt powerless to help. I felt that on a smaller scale, even being only 25 miles away. I was worried about many friends and colleagues, David among them."

Greg coordinated with Iain Campbell, a JET (Japan Exchange and Teaching) program advisor, who quickly set up a Twitter account with the help of others in the local JET community and abroad. Iain's team used hashtags to represent blocks on a map indicating where people lived. Iain also set up a Google document spreadsheet asking people if they had electricity and needed food, supplies, and water. They were able to get help to many and also contact their families.

Survival was compounded by radiation fears that were pumped by sensationalized media reports. He saw headlines that Sendai had been annihilated. He wanted to try and calm people down, so he asked his father for help. A nuclear engineer who designed safety structures for nuclear power plants, his father wrote an e-mail letter in layman's terms explaining the conditions. Greg sent out the letter to everyone he could. "Sendai is about 62 miles from the Fukushima plant, a completely safe distance as far as radioactivity emitted directly from the melted fuel goes," wrote his father. "Furthermore, there are mountains between the plant and Sendai, and the wind was blowing out to sea when the greatest amount of radioactive material was expelled from the exploded reactors. It's very unlikely that this would pose a threat to Sendai."

After a few weeks when things started to stabilize and gasoline was available, Greg and a few friends volunteered for cleanup in Sendai, Ishinomaki, and Kesennuma. Some of their work involved shoveling out mud from flooded apartment buildings and removing wrecked furniture.

In early April, with school delayed for a month, Greg established an NPO with the help of Kyle Maclachan, a friend and fellow ALT. Kyle used to be an emergency worker in Virginia and had a lot of experience with NPO setup and submissions. It made sense to collaborate with a preexisting NPO rather than start from scratch. The Miyagi English Education Support Association (MEESA) agreed to work with them. Greg set up a YouTube channel and Flickr account, where they uploaded videos and photos of volunteer activity. They were also able to collect a bit of money. The focus was education, but it was difficult to find something not covered by another group or involving a lot of red tape. MEESA had a contact with a Japanese stationery company that sold them school supplies at cost. They ended up buying the supplies and donating to schools and students in need.

By May, when school resumed, Greg was surprised at the way people just wanted to move forward and not dwell on the tragedies. "One of the strangest things for me was the realization that everyday life doesn't stop after a disaster, the way it does in a movie. You still need to worry about who'll get the food, who'll cook it, and who'll do the dishes," he says. Greg felt like he was living a parallel life. Part of him was in a cataclysmic disaster in Japan, and part of him was dealing with just day-to-day things. The disparate levels of impact also amazed him. Some people lost everything. Others were not affected at all. Sometimes it was the difference of one city block. Greg only lost a few plates that fell off his dish rack. There were others, like David, who had a near-death experience. "It was uncomfortable for people like me to be put in the same category of 'tsunami survivors,'" he says. "Not everyone understood the complexity of this disaster."

He reveals a sense of lingering guilt. "Survivor guilt isn't the right word for it, but a feeling that you hope you did all that you could," he says with a subtle frown. "David's school in Nobiru didn't get a tsunami warning. No one knows why." The warnings, he admits, were sort of a running joke among those living and working in the coastal region, and they were not taken seriously. "This was the one time the warnings could have saved people. It really upset me at the time," he reveals. He feels that he should have sent David a text message saying the tsunami was coming. He did not know the message could not get through or that the school had not heard the tsunami warning. "Still, I should have sent it," he says. "Those thoughts still go through my head."

DAVID'S FAMILY AND FRIENDS were so relieved to see him when he returned to the States. They fed him lots of good food immediately upon his return as he had lost quite a bit of weight. By this time everyone had seen his *60 Minutes* interview but asked him to tell his rescue story anyway. "Are you really sure you want to go back?" they asked. His friends and siblings could understand why he wanted to return, but his parents were very concerned about the radiation and pressured him to stay.

David felt that his near-death experience had a purpose and that he needed to return. "It's as if God has kept me alive for a reason," he revealed to his parents. "I think people need my help." In this light, his return made sense to them. "If that's your feeling," his father said, "you should go back."

EIGHT

Help Us, Please!

Is my life really going to end this way?

—*Toru Saito*

I T HAD TO BE A MIRACLE. TORU WAS SURE HIS FATHER HAD
been swept away by the tsunami that pummeled their seaside
village of Oginohama just four hours before. Toru had last seen
him heading for the village harbor. As a volunteer fireman, it
was his father's job to close the seawall floodgates after a quake. The
risk of a tsunami was always high. The many past tsunamis along the
region's Sanriku coastline had swept away their villagers and thousands of others before them.

But here he was, safe with his family at the evacuation site, Toru's former junior high school. They were alive, but the waves had
destroyed their home and lumber factory, both located near the sea.
Their livelihood and generations of memories were wiped out in only
a few hours. Like all of the 100 or so villagers with them, they huddled for warmth in the dark, tending to the elderly and children. The
reality of their circumstances was becoming clear as they sat soaked,
freezing, and in shock. With electricity down, there was no light or
heat. They would have to search for water and food. They were left
isolated, as roads to the nearest town were blocked by tons of debris
left by the tsunami. Their idyllic seaside villages were nearly all ravaged. Fate and the formation of certain inlets had saved a few.

Toru looked across the playground out to the familiar sea below.
The beach where he had often played with classmates was now gone.
Shivering, he tried to remember the warmth of the many golden sun-
sets he had seen there and the sparkling light reflected from the water,
soothing and hypnotic. But all he could feel was a terrible emptiness, cloaked by an unimaginable nightmare. At just 18, he had lost
nearly everything. He worried if he would ever be able to go to the

university. "Is my life really going to end this way?" he thought, glancing upward. The bitter cold snow, falling since their tsunami escape, had stopped and cleared, revealing a vast open sky filled with glistening stars. For a moment it lifted his spirits. *Surely help will come,* he thought. *But when?*

In the coming days they were able to survive with fresh water sourced from mountain streams. Food, blankets, and bedding were gathered from three houses still intact in nearby Samurai Village. But there were elderly and others among the group who needed care. They would have to get help by air since the roads were all blocked. They could only hope that rescuers in a helicopter would see them. Using the traditional red-and-white celebratory cloth prepared for school graduation, they laid out a big SOS on the playground.

On Sunday, March 13, their prayers were answered when they heard the loud mechanical whirring of a Japan Self-Defense Forces helicopter. It had been difficult for the airmen to spot the junior high evacuation site from above, tucked in a craggy, narrow inlet. With helicopters limited and hundreds stranded needing emergency help, it was taking time. It took two attempts before the feeble elderly were convinced to board the aircraft. An exhausted mother and her hungry infant joined them. The others were left to fend for themselves.

With a fisherman's knack for survival, the villagers were able to manage. Toilets were fitted with a flushing system, and a kerosene stove became the main source of light. Fishing gear and tools were used to refashion pots and tanks meant for seaweed and fish. They hand built a water storage tank, cooking pots, and a bath. "Our bath was the first one built, and other villagers came for a look," says Toru's mother proudly. "We surprised ourselves with the way we came up with that survival know-how."

The tsunami-ravaged towns up and down the Sanriku Coast were faced with the same daunting challenge of basic survival. In Tohoku's many small communities, populated with generations of connected families, relations are close-knit. Mutual support and cooperation are the norm. The evacuation sites became a mini composite of the town itself, with the role of command already clear. But at evacuation centers in cities such as Mayor Sakurai's Minamisoma, about 62 miles to the south, many residents were strangers. Roles were undefined and relations strained to the limit. Leaders emerged to help manage day-to-day needs, but the unprecedented triple disaster of quake, tsunami, and nuclear accident created insurmountable stress that often threw cooperation to the wind.

A Self-Defense Forces unit was able to reach Toru and the villagers by land about a week later after they had managed, with backbreaking effort, to clear away some of the road debris. The rations and supplies they brought were a welcome relief, a sentiment echoed throughout the region. With their foliage-green helmets and uniforms, the SDF soldiers became a common sight around Tohoku in a country that had, until then, been deeply ambivalent about the presence of its defense forces. Before the quake, men in military uniform were rarely seen outside of their bases, except during annual disaster drills. Now in the aftermath, the Japanese soldiers were the first to respond. They searched for dead bodies, digging through high mounds of debris and mud often soaked with toxic chemicals from destroyed factories. They pulled wrecked boats and cars off of buildings, sprayed water into the crippled reactors at the Fukushima plant, and performed every other rescue and recovery feat imaginable.

Over 100,000 of the country's 240,000 soldiers were dispatched to the hard-hit areas, the biggest deployment of SDF personnel in postwar Japan. It was five times larger than the number sent to Kobe after the 1995 earthquake there. By mid-April they had rescued or assisted 19,000 civilians.[1] The soldiers' efforts won wide praise throughout Japan.

The role of the Self-Defense Forces is bound up with the country's wartime history and its aftermath. Japan's military forces are framed by the legal architecture drawn up during the US-led occupation of the country from 1945 to 1952. The constitution's famous Article 9 states that Japan must not use force to resolve international disputes, banning it from making a military designed for war. The aim of the occupiers was to prevent a repeat of Japan's aggressive militarism and to limit the power of the monarchy, though the clause found huge support among ordinary Japanese, too.

Japan's military expenditures throughout most of the postwar period hovered at around 1 percent of the GDP as the nation sheltered under the US defense umbrella. It is small, but still a hefty chunk as the world's third biggest economy. But interpretations on the terms *peacekeeping* and *defense* are being tested and stretched, especially in light of China's growing military presence in Asia. As territorial conflicts with China and threats from North Korea have increased, so have SDF activities, pushed by a series of nationalist prime ministers.

Japan's military security still comes largely from its close ties with the United States. Just after World War II, the United States became Japan's military administrator, formalized with Japan's signing of the Japanese Instrument of Surrender on September 2, 1945. Under US occupation, the shattered country was able to rebuild with guaranteed national security. The 1951 Japan-US Security Treaty enabled the United States to maintain a strategic military presence in the Western Pacific. This was cemented with the revised Japan-US Security Treaty signed in 1960. Ultimately, however, the financial cost of that security has been shouldered largely by Japan. In December 2010, the two nations agreed to a recalculated amount of US$2.3 billion that Japan would pay annually for a five-year period from fiscal year 2011.[2] This issue and long-term clashes over US military bases in Okinawa have sparked debate over restructuring the alliance and the need for Japan to establish greater military independence.

The March 11 quake and tsunami became a catalyst for reconciliation as Japan struggled with rescue efforts in the immediate aftermath of the disaster. The United States offered a helping hand, initiated by the US Department of Defense, with a newly formed, bilateral emergency disaster relief effort called Operation Tomodachi (Operation Friend).

For the islanders of Oshima, the US Marines with Operation Tomodachi were a sight for sore eyes when they arrived two weeks after the quake and tsunami. The island's 3,400 residents, the largest group in the region, had been left isolated. They had no electricity, had limited communication capability, and were rationing fresh water and food. SDF helicopter drops had provided some supplies and assistance, but a larger effort was needed.

Located just south of Rikuzentakata, the island was normally a 25-minute boat ride from the mainland port of Kesennuma and accessible by one bridge. The bridge was wiped out by the tsunami, and rescue efforts were thwarted by fire and a raised seabed that prevented ferries and emergency boats from docking at the island's main port.

The fire had burned dangerously out of control. The crushing waves had cracked open petroleum storage tanks that left a putrid, toxic spill. The slick had burst into flames, devouring partially submerged houses and spreading to buildings and fish factories on the mainland in nearby Kesennuma. The undersea quake occurred at a relatively shallow depth of about 15 miles, releasing most of the energy at the seafloor. The massive tremors that followed were the result of a violent uplift of the seafloor propelled by tectonic plate movement. This action not only left the seabed elevated or dropped along the Tohoku coastline, it moved Japan's main island of Honshu a reported 8 feet.[3]

These factors, combined with masses of debris, made access impossible for ferries and rescue boats that could deliver emergency supplies and equipment. The SDF did not have the flat-bottom landing craft needed to dock in shallow water and carry large vehicles and aid supplies, so they turned to Operation Tomodachi for help.

Lieutenant Karl Hendler and Corporal Kevin Miller of the Marine Corps were among the troops who came to Oshima Island's aid as part of the operation. When they first heard the news of Japan's quake and tsunami, they were in the Malaysian port of Kota Kinabalu. They had pulled in a few hours before on the USS *Essex* as part of the 31st Marine Expeditionary Unit (MEU) and had just checked into their hotel when they received the call to come back to the ship right away.[4]

Once the USS *Essex* reached its Oshima Island destination on April 3, marines were sent ashore to liaise with the Japanese forces and other American units to assess what could be done. Several companies from the 31st MEU infantry battalion were sent in, but they realized they needed more help. As a member of the MEU command element, Lieutenant Hendler was then assigned to lead a 42-member platoon that included Corporal Miller. Their destination was the small fishing village of Kamagata at the southern end of the island where they would hike after landing.

From the USS *Essex* to the island, they were transported by landing craft stacked high with machinery and relief supplies. The marines huddled in the cold wind during the 12-mile, one-hour trip. As they approached the shoreline, they were shocked by the sight. Pictures they had seen from the news media during their weeklong transit had not revealed the extent of the damage.

The tsunami had destroyed nearly everything it had touched. Massive petroleum tanks, punctured and crumpled, lay sideways like beached, dead whales. Wooden buildings and homes were shattered and cement structures gutted, leaving mounds of splintered planks and rubble. Wrecked cars lay scattered about. Many small and large fishing boats were washed ashore, some dragged up the island's famous forested hillsides. The three-square-mile island had been dubbed the "Green Pearl" and a portion designated a national park. Now its shoreline, ports, and fishing villages were a brown-tinted debris-filled wasteland.

"When we got there, I realized the pictures didn't do it justice," says Lieutenant Hendler.[5] The 25-year-old officer from Cleveland, Ohio, had begun his military career about two years before. "It seemed like we could be there for a hundred years just moving debris," he says. Corporal Miller had also never seen devastation of that magnitude. "But one of the things marines are good at is getting an order and immediately executing," he says. "And that's what we did."

As the column of 43 marines hiked toward Kamagata Village along the half-frozen muddy roadways, local residents would stop and bow repeatedly as they passed by. "You wouldn't expect that type of formal greeting because they were in the middle of working themselves," says the lieutenant. "They were all carrying heavy loads and water back to their houses."

A gift of candies and a thank-you letter also made a deep impression. Seemingly out of nowhere, a man appeared and handed the lieutenant a bag of candy and bowed. It was a humbling moment for the group, knowing that the man probably did not have much food. One young woman also conveyed her appreciation with a thank-you note, written in English and Japanese. "We're just used to working, working, working and rarely get to see appreciation for the results or efforts," he says. "The marines in the platoon could see what we were doing was real important."

At the fishing village, the marines began clearing the small port. They lifted boats back into the water and removed mounds of wooden planks and debris from destroyed homes nearby. That evening they set up camp near the SDF in the hills of the national park while snow flurries drifted through tree boughs and swept across the frozen ground. They had been warned about the weather and brought every piece of warm clothing they could. Local residents praised the marines and SDF soldiers for toughing it out in tents while national police assigned to the emergency slept in the hotel nearby.

During the second day, many of the village residents came to join them. The marines were pleasantly surprised, as they rarely get an opportunity to interact with civilians and are used to working only among themselves. Corporal Miller recalls one colleague saying, "'A local just waved at me and I don't know what to do.' I said to him, 'Well, just wave back!'"

The 20-year-old corporal from Knoxville, Tennessee, was a musician before he joined the marines two and a half years before. He had traveled a lot for performance gigs and as the son of a military man but wanted to experience more. Knowing his passion for helping others, a relative encouraged him to join the MEU, which serves in humanitarian assistance, disaster relief, and evacuation operations. For the corporal, helping others has always been an integral part of his life.

A captain from the SDF gave the platoon guidance using hand gestures, as neither side spoke the other's language. The captain explained that within the rubble there were many personal items that they wanted saved—things that seemed to have sentimental value such as photo albums, pictures, books, and little knickknacks. Everything else was going to be thrown away. The marines made a pile of a few hundred little items in hopes that the residents would get them when they came back to the village. "That's what made it the most real for us," says Lieutenant Hendler. "It wasn't just trash we were picking up. It was people's whole lives."

The Oshima Island residents say they will never forget the marines' help. Eight-year-old islander Wataru Kikuta calls them his heroes. "I made with them a 'bridge for tomorrow,'" he says shyly while describing the little dam he made with the marines' help. His mother, Reiko, shows off a Marine Corps pin he received as thanks for his help. "We also called it 'bridge over troubled water,'" she adds with a smile. Wataru wanted to join the Marine Corps island cleanup after watching them hoist and carry his family's heavy fishing boat back to the sea after it had been swept far inland by the tsunami.[6]

Wataru, his parents, and his grandparents living with them had barely survived the tsunami onslaught. Fortunately, Wataru was at school when the tsunami hit. Just after the quake, his mother ran to the school, located on higher ground, to make sure he stayed there safe with the other students and teachers. Reiko knew well the potential dangers after hearing stories from her mother, who lived through the 1960 Chile tsunami.

Wataru's grandparents, in the house at the time, also needed to be in a safe location. His grandmother was able to go to a friend's home, but his grandfather was unable to walk. His leg was in a cast after a recent accident. Reiko and her husband were able to take him to a relative's home just in time. The tsunami destroyed their house next to the bay, while waves reached within just a few feet of their fish warehouse located farther back. They were now struggling to manage day to day while living in the warehouse.

Wataru was able to join the island's Operation Tomodachi effort with some guidance from his father and the friendly marines at the cleanup site in his village. As he was building the tiny dam across a roadway, a Japanese news photographer and TV cameraman spotted the collaboration. Wataru appeared the next day on national TV and in newspapers across the country. He became a virtual poster child for the marines' mission. His mother says that the marines lifted their spirits and inspired them to begin restoring their lives. "It was also a great educational experience for Wataru," she says. "He'll carry that for the rest of his life."

Operation Tomodachi helped boost the image of US forces in Japan. Surveys found that opinion in Japan toward the United States following the operation was the most favorable in nearly a decade. From March 12 to about May 11, working under guidance from the SDF, the operation deployed 24 US naval ships, 189 aircraft, and almost 24,000 US service members. Both countries have reportedly shared the $90 million cost.[7]

During the first week after the quake and tsunami, US forces rescued about 20,000 people and restored transportation facilities including Sendai Airport, a vital air hub for the region. The operation was also an important asset during the early stages of the Fukushima Daiichi plant accident. The US Navy provided almost two million liters of fresh water to cool the plant's reactors, and the Marine Corps Chemical Biological Incident Response Force trained SDF troops operating nearby. US unmanned aerial drones flew over the plant to monitor and collect data for the Japanese government. On-the-ground assistance was provided by officials from the NRC and the US Defense and Energy Departments.[8]

Called "the single largest humanitarian relief effort in American history," the peacetime mission has been lauded by both governments and mainstream press as a great success. But there have been dissenting opinions and issues between the Japan and US sides, mainly on the lack of effective and classified communication systems, vital during regional tensions or in a time of war.[9]

There are still 91 US military bases in Japan, with 38 located in the Okinawa Islands. On the main island, bases occupy 19 percent of choice land. For 27 years after World War II, Okinawa was under US military administration. May 15, 2012, marked the islands' fortieth anniversary of reversion to Japan. Since reversion, there have been vociferous and angry calls for the US military to withdraw from the islands. An agreement between the United States and Japan in April 2012 will lead to about 9,000, or half, of the United States' troops withdrawing from the islands to Guam and other locations in the Asia-Pacific.[10] Okinawa's location near China, and concerns over its growing military presence in the region, will likely ensure that the US military remains a permanent fixture there.

THE MARINES AND A WAVE OF INTERNATIONAL RESCUE TEAMS became the symbiotic inward flow of foreigners as so-called *flyjin* were leaving. A 15-member emergency team from China and teams with search dogs from the United States, Germany, and Switzerland were among the first to arrive on Monday, March 14. As of March 29, there were 25 teams from countries worldwide offering humanitarian aid and supplies. One hundred thirty-four governments and 39 international organizations had offered assistance. Nearly one year after the disaster, it was estimated that donations had exceeded US$6.4 billion (¥520 billion), and 930,000 people had been involved in volunteer work.[11]

Actors, singers, and celebrities also offered their support. Japanese actor Ken Watanabe organized a website called Kizuna 311 with video messages of encouragement from Hollywood friends. In one video, Watanabe is shown reading aloud Kenji Miyazawa's poem "Strong in the Rain." Asian singers and entertainers, including Jackie Chan, also produced a "Strong in the Rain"–themed song of support.

The music scene in Japan ground to a halt in the days following the quake and tsunami, but there were a few stalwarts. Singer-songwriter Jack Johnson was in his hotel room in Osaka when the earthquake hit, but he went ahead with his scheduled concert just hours later as TV images of the tsunami flashed across screens worldwide. He had to discontinue his planned tour but later donated $50,000 to a Japan relief fund. Pop singer Cyndi Lauper cheered audiences during her scheduled concert tour after not fleeing when she arrived the day of the disaster. She has since visited the Tohoku disaster zone and become an unofficial spokesperson for the region.[12]

JAPAN HAS BEEN ONE OF THE WORLD'S most generous donators of emergency aid and money to nations in need during disasters. But it took the government many mistakes to welcome international aid within Japan. Painful lessons had been learned from the 1995 Kobe earthquake when government agencies proudly refused help from abroad, saying Japan could manage on its own. Victims trapped under crushed buildings or who perished in fires that followed could have been saved by the international search-and-rescue teams offering help. That mistake kick-started a new kind of volunteer movement in Japan, along with increasing interest in nonprofits and nongovernmental organizations, both domestic and international.

At the hub of this effort for over 25 years is Sarajean Rossitto. An American based in Tokyo, she works as a nonprofit NGO (nongovernmental organization) consultant. A large part of her work includes facilitating workshops, seminars, and training programs that teach the how-tos of running and working with nonprofits. She also advises on best-practice partnerships for a wide range of clients from local communities to Japanese and international corporations. During the Tohoku disasters, she became a vital conduit for information

and advice on ways to donate, offer help, and work with rescue organizations. Her blog, updated constantly during those first few weeks,
was a beacon of light in a dark, desperate time.[13]

Almost immediately after the quake and tsunami, she began receiving a flood of e-mails from people both in Japan and overseas
offering donations and asking who would take volunteers. Since she
was not prepared to go up to Tohoku, she decided to start a website that could answer questions and connect people with rescue efforts. From the evening of Saturday, March 12, through the next two
weeks, she was online about 15 hours a day. During the day, she sent
out volunteer information via Twitter and Facebook that included
a link to her website. At night, she answered e-mails from overseas.

Sarajean quickly learned that the March 11 disaster had un-
earthed a whole new set of challenges for nonprofit and NGO aid
efforts in Japan. One of the most startling was the discovery that
Japan, although a quake- and tsunami-prone country, had no
government-run disaster coordination mechanism for nonprofits and
NGOs to use domestically. The government only had a system for
funding and coordinating activities for these organizations to use in
disasters abroad. The irony was clear, especially when the Ministry
of Foreign Affairs was willing to use its disaster support network to
start the domestic network. But setting up the channels to create and
allocate a domestic disaster fund for nonprofits and NGOs became a
logistics nightmare.

Sarajean also cites the great divide between local administrators in the hard-hit towns and the many eager volunteers offering
help. Local municipalities across Japan had previously established
disaster volunteer centers to coordinate relief efforts should a disaster or major emergency occur. Volunteer groups were required to
first consult with the center before embarking on aid activity in the
area. The concept was a good one, but towns were overwhelmed

by the severity of the March 11 disaster, and often centers did not coordinate well. There were many volunteer groups, both Japanese and non-Japanese, that effectively bypassed the center system and concentrated on the small, isolated towns that were being ignored by the bigger organizations.

Some local centers, despite good intentions, were out of sync with residents' needs, while others were clearly progressive. One local disaster center explained that it was sending out notices to residents via the Internet, asking who needed help. But most of the survivors and affected people did not have Internet access. Many were elderly who did not use the Internet. Sarajean cites the city of Ishinomaki as a successful example. "The mayor was very smart and wanted to strategize with NPOs on coordinating help."

Radiation from the Fukushima accident became a major issue for disaster relief organizations, domestic and international. The few that chose to work in the irradiated evacuation zone were not officially recognized. Animal rescue groups, such as JEARS (Japan Earthquake Animal Rescue and Support) and ARK (Animal Rescue in Kansai), made heroic efforts to save abandoned pets and farm animals. Many were left to starve to death or fend for themselves. Among the disasters' shocking images were heartbreaking photos of emaciated horses or dogs tethered tightly to posts, unable to escape. In the hierarchy of needs, unfortunately animal rescue was not seen as a high issue.

The Japanese Red Cross drew a swath of complaints. It was the biggest philanthropy distributing funds in the stricken areas, its donations coming to a hefty US$1 billion. But the organization came under fire for being too slow on distributions. "People were complaining, but they don't understand how it works," says Sarajean. "The Red Cross is a facilitator of a political system. The whole cash donation system needs to be divided through a political process."[14]

ONE INTERNATIONAL NGO offering long-term and innovative economic support has been Oregon-based Mercy Corps. Randy Martin,
the former director of Mercy Corps' global emergency operations,
came to Japan for about two years to strategize and monitor economic recovery efforts and funding with partners Peace Winds Japan
and PlaNet Finance Japan. An Illinois native from an agro-industrial
town bordered by a sea of soybean fields, Randy first took an interest
in international work at a young age. "I didn't hear a foreign language until I left my hometown," he reveals. He got hooked during
a college program in Micronesia and is now a veteran of some of the
world's most severe disasters. The scars are not apparent. Instead, a
kind but penetrating gaze reflects a life dedicated to helping those in
need.[15]

Malka Older, with Mercy Corps at the time of the disaster,
quickly followed Randy to the disaster zone during the second week.
She was eventually seconded to Peace Winds Japan as their on-the-
ground economic recovery manager. An organizational whiz, fluent
in Japanese, she had previously lived on the southern island of Kyushu while in the JET program. By the time she was 30, she already
had considerable disaster experience, including the catastrophic
2004 Indian Ocean tsunami.[16]

When she woke up on March 11 and saw the news, she hoped it
was not as bad as it sounded. In her mind she had filed the Southeast
Asia tsunami as a once-in-a-lifetime experience. She admits she was
skeptical about coming to Japan to help. Why would an NGO want
to work in such a wealthy country? But in a disaster, even wealthy
countries need technical assistance. Organization and coordination,
she says, are often more important than money. Malka was able to
assist with two important programs.

One involved vouchers to replace inaccessible cash. Local bank
branches had either been destroyed by the tsunami or become

inoperable. Credit cards were unusable locally. The vouchers would allow residents to buy the items they needed and not depend upon donations. "Economic recovery comes with an injection of cash," she says. This would keep businesses running and local staff employed. It was a simple but effective way to revive the communities using the local economy, instead of an expensive relief operation.

It was also in the interest of efficiency and distribution. Distributing items became a major issue at evacuation centers. Out of fairness, the same item was given to everyone or no one at all. Unless there were enough coats for everyone, no one would get one. There were lots of donated items thrown out or left in storage.

Mercy Corps received more than $6 million in donations for use in Japan and concentrated on the devastated towns of Rikuzentakata, Ofunato, Kesennuma, and Minamisanriku. "Every disaster is remarkably different," says Randy. "I don't think I've ever seen that much debris, or such a fast response." He was amazed that almost everyone got into a shelter within 48 hours, despite the chaos. In the 25 or 30 years he has been doing this work, it was the first time he had seen it done so quickly. "It was unprecedented," he says.

The response also felt different for Malka. "The Self-Defense Force army did an amazing job," she says. "Drinking water, baths, laundry trucks with washing machines. They were really walking the walk." There was an unusual absence of NGOs because they needed to be registered in Japan beforehand. Mercy Corps was allowed in because of its relationship with Japan-based Peace Winds. The Japanese government handled coordination with NGOs. In an undeveloped country, the United Nations would be the coordinator.

During his first day in Tohoku, Randy was brought to Kesennuma by a helicopter carrying a load of oranges for disaster relief. After landing, he went to a store that was open for business, and the first thing he saw inside the door was oranges. It was a reminder that often what is distributed is not what is needed. "It undermines market recovery and makes the local people dependent on recovery funds longer," he says. Their initial goal was to gradually move away from distributing relief goods to market-based support.

After the first week, stores were beginning to reopen, and the commercial supply chain was coming back. Randy was drawn to a convenience store in Kesennuma where the tsunami had gone right through it. Not only had they cleaned up debris, scooped out the mud, had their staff in uniform, and were selling items at discount, they also had an employee outside holding a big sign saying, "WE'RE OPEN." They even had fresh rice balls.

A fish processor surrounded by a Grand Canyon of debris about 20 feet high on either side also caught his eye. The processing building did not get swept away, and the upper floor housed the generators. They had already cleared all the debris and were ready for the boats to come in.

Another big challenge for residents was mobility after almost everyone lost their vehicles. One Peace Winds program, supported by Mercy Corps, offered a bus service from the evacuation centers to shops and other locations. In towns like Rikuzentakata that were almost totally destroyed, the bus service would go to neighboring towns such as Ofunato where shops were operating.

For merchants who lost their stores, Mercy Corps and Peace Winds helped them get mobile with minishops on the back of trucks. The mobile shops first went to evacuation centers and then to temporary housing units. It was especially helpful for the elderly. But Malka notes that "if it was Indonesia, there would have been a dozen shops set up outside the evacuation centers right away." She was surprised at the lack of entrepreneurialism and reluctance to try new things. At the start, the local chamber of commerce was encouraging and pleased with the idea, but very concerned about regulations and licensing, despite the circumstances. There had to be the right kind of vehicle for the right kind of food.

She and her team went to the town hall to expedite things but were faced with more of the same. It was a puzzling lack of flexibility. One merchant who lost his shop finally agreed to take the plunge. "I'm just going to go ahead and do it!" he told them. Later, they were surprised to hear him ask, "Will the people in America be happy if they know I'm making a profit?" He was concerned that the aid he received was being used to turn a profit rather than using the assistance to offer cheaper food. Malka explained to him that self-sufficiency was the goal. They were trying to restart the economy and hoped to get the mobile trucks working for profit rather than supporting the community through additional donations. Eventually the idea caught on and more vendors participated. Everyone was happy when the trucks came by, especially the elderly and those without transportation who were living in the temporary housing in remote sites.

After "phase one" disaster relief, "phase two" became recovery and rebuilding. But hard-hit businesses were reluctant to take out loans to rebuild. Some owners were burdened with overlapping debt such as home loans. The prefectural and national governments were offering large grants and matching support, but reimbursement would not come until 2012. Fisheries were in a bind because of their seasonal nature. In Ofunato their popular catch of mackerel pike would be in September. It was sold nationally and a major source of income for the year. They needed funds right away to borrow a building and source an array of equipment such as forklifts, crates, and fish tanks.

Microfinancing was a viable answer. Mercy Corps, in a tie-up with PlaNet Finance Japan and the Kesennuma Shinkin Bank, designed a program to provide microfinancing for hard-hit small businesses with fewer than 20 employees, which they called "Sanriku Tomodachi Fund for Economic Recovery." PlaNet Finance Japan is a federated partner of Planet Finance Group, a French NGO that focuses on microfinance programming in developing countries. Shinkin Bank is a national network of banks in Japan that works like a cooperative bank or credit union and offers loans to small businesses. They had lost 7 of their 12 branches in the area. Mercy Corps had a little over $4 million in donations in addition to the Shinkin Bank capital. It was unusual for NGOs to work with a bank.

The program has been spotlighted in the media for its success and is now being used as a model for business recovery in the region. "One of the beauties of the program is it's easy to scale up," says Randy. "It also becomes more efficient the more people give as operating costs do not increase."

MAYOR SAKURAI'S PLEA FOR HELP rang out across the world. Blankets, clothes, and food began arriving in Japan, addressed to Minamisoma from America, Europe, and many other countries. The deliveries were sent to the Soma port or other areas away from the radiation zone, and staff from the city office went to pick them up. Some people sent dosimeters and other equipment for measuring radiation. After a few months, the city had to ask that the aid be stopped. They did not know what to do with all the boxes of supplies.

Setsuko credits UNICEF with aiding and restarting the Rikuzentakata nursery school where she worked as a cook. The organization's staff and volunteers listened carefully to their needs. They helped clean and paint the new borrowed facility and donated many supplies. Setsuko says everyone was extremely grateful for UNICEF's help. "We could not have done it without them."

NINE

Departures

Hope is the key.

—*Alfons Deeken*

SETSUKO JUST HAD TO FIND HER HUSBAND, TAKUYA. IN her heart, she still held hope that he was alive. But a dreaded thought was sinking her resolve. *Did the tsunami take him?* she agonized while walking quickly toward the city's temporary emergency headquarters on the morning of the twelfth. It was a new building where meals were normally prepared for public schools and facilities, and it was set on a high location. *Surely he is there. Probably organizing efforts to help survivors,* she thought. *He's a born leader and always ready to help others.*

It had been a sleepless, freezing night caring for the remaining nursery school children after their near escape from the tsunami just 16 hours before. Relieved parents who found them had shared their terrifying stories. Setsuko was bracing herself for the worst. A mix of fear, worry, and love pushed her onward. But she was unprepared for the unfolding tragedies ahead as she stepped into the familiar building.

Hundreds of survivors, cold, muddied, and in shock, were desperately seeking word of their families. Tables were being set up with volunteers to help with a growing stream of inquiries. So many questions could not be answered. The tsunami had wiped out the city hall and all records of the town's citizens. It was as if no one in Rikuzentakata had ever existed.

Futoshi Toba, the mayor elected just one month before, and his surviving staff had formed an emergency coalition with the police and fire department. Slowly and painfully, organization was emerging from the chaos despite often unbearable personal tragedy. The mayor's wife, Kumi, had been swept away by the tsunami from their home closer to the sea. As he escaped to the roof of the city hall, he considered rushing to their home to save her. But his duties as mayor held him back. His colleagues and the city needed him. His two sons, ages 10 and 12, had survived at their school on higher ground.[1]

For Setsuko, it was becoming agonizingly clear that many government workers had been swept away. She heard that the staff had evacuated from the city hall after the quake. *Was it possible Takuya stayed on the ground to help those in need?* she wondered. Setsuko could feel her heart racing as she searched for her husband among the throng. But there was no sign of him. She realized that the best way to get word of him was to volunteer for the missing-persons desk.

With electricity out and computers down, she had to write everyone's name, contact information, and description of their missing relatives. But as more and more people poured into the center, she was overwhelmed. Long lines began to form as the day wore on. "How can I find my mother . . . my father . . . my grandparents . . . my siblings . . . my child?" people would ask in tears, anger, disbelief, and frustration. Some were Setsuko's close friends and neighbors. It was all the more difficult because she knew exactly how they felt. The only consolation was that everyone had been affected in some way. No one was alone in his or her trauma, grief, and suffering.

That night, exhausted beyond measure, Setsuko and her colleague from the nursery school cried themselves to sleep as they huddled together under a single blanket for warmth and comfort. The woman's husband had survived, but her two young children and parents were still missing. She was sure they had been swept away by the tsunami as they tried to escape by car. Setsuko had heard the speculation repeated many times throughout the day. Drivers and passengers were engulfed after their cars were caught in traffic jams when everyone rushed to escape. Sometimes, the difference between life and death was one single turn onto the right or wrong road. *If only people had run instead,* she thought. *If only the evacuation sites had been chosen more carefully, above the tsunami demarcation line. If only the city hadn't allowed buildings to be so close to the sea.*[2]

The next day, Setsuko joined two other volunteers at the death registration table. It was here that family members would come after they had identified their relative's body at one of several public school gymnasiums that had been turned into temporary morgues. Again, she and the other volunteers faced long lines of grief-stricken people. Along with filling out makeshift forms, they had to help arrange cremations.

Rikuzentakata's crematorium was partially damaged and could only burn about seven bodies a day. With 700 bodies found so far, Mayor Toba was asking crematoriums in neighboring towns for assistance to avoid a mass grave.[3]

For many towns, like Higashi-Matsushima where David Chumreonlert lived, there were not enough cremation facilities available. Families had to contend with temporary mass burial in dirt pits. Burials in Japan were the norm before and during World War II. The custom of cremation spread in parallel with rapid postwar economic growth and urbanization. Citing limited land and sanitary reasons, large metropolises such as Tokyo and Osaka require cremation. The emperor and empress broke with imperial burial tradition when in 2012 they indicated a preference for cremation.[4]

At the death registry, Setsuko and the other volunteers had to explain the difficult procedure. Family members needed to put the corpse on a burnable wooden plank, wrap it in a blanket, and then find a way to transport it to their assigned crematorium. But finding a plank and spare blanket was nearly impossible since most families had lost their home. A lack of vehicles and a gasoline shortage compounded the difficulties. Coffins and urns for cremated remains were also in short supply because several of the town's undertakers had been swept away by the tsunami.

The bereaved were painfully conflicted over their love and affection for the deceased and revulsion toward the corpses, some of which were badly damaged. Some people were outraged at the request. Setsuko tried her best to gently remind them that they were all in the same situation.

THE WORK OF HANDLING DEAD BODIES in Japan now goes mainly to hospital nurses and funeral parlors, yet the practice is not wholly institutionalized. There is still reverence for the corpse and the belief that it must be cared for with proper ritual in order for the spirit to transition to "Paradise" or the "Pure Land" and "Buddhahood." This belief, born from Buddhism and Shintoism,[5] was strongly evident after the 2011 tsunami during the long, gruesome search for the missing. Some continued to search for loved ones more than a year later.[6]

Rituals surrounding death have been modified as Japan has modernized. But in rural regions such as Tohoku, where time drifts slowly, traditions remain. Until the 1950s, most deaths occurred at home, and families would tend to the corpse. With improved health care, hospitals have become the common place to die.

Nearly 85 percent of Japanese funerals are Buddhist, with funerary rites influenced by sect and region. It is usually a two- or three-day process beginning with a wake (*otsuya*). This is typically held the first evening after death at the family home or the next day at a funeral parlor (*sōgisha*). If at home, the body is first laid on a futon and covered with a sheet, with viewing by relatives and neighbors. A family member stays with the corpse throughout the night as "protector" and to confirm that death has actually occurred. A mortician (*nōkanshi*) dresses the body and lays it into a coffin for viewing by more guests. Food and drink are served at the wake, known to turn the solemn event into a drunken revelry.

Embalming is not customary, and dry ice is used to preserve the corpse until cremation. Sometimes items such as *zōri* (Japanese straw sandals), gloves, or walking sticks are added to the coffin to assist the dead on their journey to the next world. The funeral service (*osōshiki*) is usually held the following day, at either a funeral parlor or Buddhist temple.

It is customary for anyone other than next of kin to offer condolence money (*kōden*) during the wake or funeral service. The amount depends upon the person's age, wealth, and relationship with the deceased and can range from ¥3,000 to ¥30,000 (about US$38 to US$380). This helps pay for the cost of the funeral, which can average about ¥2.3 million (US$29,000). The family later returns the favor with a thank-you gift about half or one-quarter the value of the condolence money.[7]

After the funeral, the deceased is transported by ornate hearse to a crematorium (*kasōba*). Family members witness the coffin slide into the burning chamber. The cremation takes about 90 minutes for an adult corpse. Afterward, in an emotional ritual, relatives use special chopsticks to pick out bone fragments from the ashes. The hyoid bone (lingual bone) above the Adam's apple is considered particularly important because the shape is similar to the Buddha's hands in prayer during meditation. The bones are placed in a cinerary urn and kept in the family Buddhist altar (*butsudan*) at home until the forty-ninth day after death. It is believed that the spirit wanders this world until that day, when it transitions to the next. The urn is then transferred to the family grave and stored in a small chamber or crypt.[8]

The deceased is given a posthumous name (*kaimyō*) to use in the afterlife.[9] It is said that the dead will look back if called by name, which slows their journey into the next world. The new name is transcribed on a memorial tablet (*ihai*), which is stored in the Buddhist altar at home. Memorial services are held on a regular basis usually determined by the family's Buddhist sect and local custom.

After March 11, morticians in the hard-hit towns were few and far between, making traditional preparation for corpses virtually impossible. In Tohoku, morticians still perform a special "encoffining" ritual in the presence of family and close friends. During the ritual, with most of the body hidden from view, a mortician gently wipes the corpse clean and carefully dresses it with great skill in a white "Buddha robe." Makeup is usually applied to the face. The mortician then places the body in a coffin.

This scene is sensitively recreated in the Japanese film *Departures*. Shinmon Aoki, a former mortician from a region near Tohoku and the inspiration for the film, cites the important connection the ritual creates between family and deceased. "It's a moment when the dead and living are communicating with each other," he explains. "There's the vertical world of the living and the horizontal world between the living and the dead. Making that horizontal connection is extremely important."[10]

Aoki's experience as a child at the end of World War II in a Manchuria refugee camp in northeastern China was the start of his long journey with death. He and his three-year-old sister got separated from their mother during the panicked exodus of Japanese from the region and were brought to the camp.[11] Just eight years old, he was unable to care for his sister, and she quickly died of malnutrition. He will never forget carrying her, tied to his back, to the camp's temporary crematorium. "This is why I have a different perception of death from most people," he explains. "I went through all the thinking process of death from my sister long ago."

When Aoki began his ten-year period as a professional mortician in 1970, male family members in Tohoku still performed the cleansing, dressing, and cremation. But this all changed, he says, with Japan's rapid economic growth, more families living apart, and increased social health insurance and hospitalization. "Until about 1955, 90 percent died at home," he estimates. "Now in cities like Tokyo, 90 percent die in the hospital and the bodies are taken directly to the crematorium." That final ritual connecting the family and deceased is fading, he explains.

It was common practice in rural communities when someone died at home for male relatives to prepare the corpse for cremation. The body was first immersed in a tub with cold water topped with hot water and then scrubbed clean. A small sword would be secured against the breast for protection against evil spirits, and the body wrapped tightly in a white "Buddha robe." The hands would be placed together in a prayer position with Buddhist prayer beads.

Aoki explains that until about 1965, the legs were folded and tied to the torso to fit into an upright, box-like coffin. For the elderly it was natural to be placed in a sitting position, as they typically had curved backs from years of bending over during rice and vegetable farming. At times, the dying would also naturally curve their body into a fetal position.

The men would carry the coffin box on their shoulders to the cremation site, a special facility with a funeral pyre on the outskirts of the community. The burning would begin in the evening and take about eight hours to complete. The change to coffins common today came after about 1965 when community cremations were moved to commercial crematoriums. Horizontal coffins fit the shape of the crematorium chamber and are easier to transport by hearse or other vehicle. But, Aoki points out with a wry smile, "It's not easy to get the corpse to lie completely flat and close the coffin top."

During his years of working intimately with death, Aoki came to recognize what he calls a "mysterious light" in the faces of those recently deceased or approaching death. "It could be that, when we are fighting one on one with death, at the bitter end we come to a point where life and death resolve themselves, and in that moment we encounter that mysterious light," he wrote. He turned to Buddhist scriptures for solace and understanding and found that this "light" was a universal phenomenon, shared across religions. He also referred to the many accounts of people "seeing a bright light" during a near-death experience.[12]

When one is at the edge of death, this sublime "light" comes, and one feels a sense of deep gratitude, forgiveness, and peace, Aoki says. If the dying person is physically strong enough, he or she will say "thank you" or communicate gratitude through his or her eyes. "In the end, they will pass away with a smile on their face, their final message," he explains. "This is why it's so important for family and loved ones to either be at their deathbed or know this truth." He cites the Kenji Miyazawa poem "Speaking with the Eyes,"[13] which he quoted during a talk for the bereaved in Minamisoma. "For the people in Tohoku, it's important they know their loved ones who were swept away in the tsunami died this way."

Speaking with the Eyes
It's not long now
It just won't let up
It gurgles and gushes
I haven't slept all night and the blood just keeps flowing
It's blue and still out there
It looks like death . . . and very soon at that
And yet I feel the most magnificent breeze!
Spring is just over my shoulder
And this clear wind rushes towards me
Swelling out of that bluest of skies

The blue is the blue of a rush mat scarred by fire

A blue forming waves in a meadow of autumn flowers

In flowers like down, in young maple leaves

Dressed in your black frock coat

You may be on your way back from a medical conference

If death takes me now I can hardly complain

Seeing how diligently and cleverly you have cared for me

Could my indifference to suffering

Despite the constant flow of blood

Be a sign that the soul has half-departed the body

My sole torment is that because of this blood

I am unable to tell you this

In your eyes I am no doubt a wretched sight

But from here . . . after all

All I can see is the clear blue sky

And a transparent wind

Aoki's words are a source of solace for many still grieving, especially those who lost children. One year after the triple disaster, mothers were still digging for the bodies of their missing children who had attended Okawa Elementary School in Ishinomaki. Several mothers even obtained a license to operate digging equipment. This was one of the tsunami's greatest tragedies: 75 children and 9 teachers and staff were swept away because of indecision over the best place to evacuate. One surviving teacher later committed suicide, overwhelmed with survivor guilt.[14]

"Even if the mothers find the bodies of their children, they will still be sad," Aoki explains. "Instead of digging, it's better for them to know that their child died smiling and left with a feeling of deep gratitude toward them, their family, and the people still living," he says. We may think that they died suffering, full of fear in the cold water all alone, but in fact, "they were able to see the clear blue sky and a transparent wind."

Finding closure, however, can be painfully elusive.[15] Some bereaved fell back on small, personal rituals to repose the soul of the dead and reconnect with those they lost. Tohoku scholar and academic Norio Akasaka speaks about a family he saw throwing flowers into the sea for relatives still missing since the tsunami. "They looked beautiful, like a scene in a movie, each one in order throwing a bouquet," he says. At the end, the family took a group photo with their backs to the ocean. "They probably didn't know how else to express their pain, sadness, and anger," he says. "I'm sure this kind of ritual has been done in many places along the Tohoku coast."

There are rumors, he says, of women seeking solace at beaches, where they go to hear the voices of their dead children and relatives. In Tohoku and other regions, the coastline is considered a kind of mystical border between this world and the next. Traditional, sacred performances are still held at beaches to welcome the spirits of the dead.

Many ghostly sightings have been reported in the media since the 2011 disaster. Akasaka cites a strange recurrence at a beach not far from Mayor Sakurai's Minamisoma. At nighttime, cars driving past the beach will suddenly hit an unseen object. Worried drivers will stop and step out of their car to look, but find nothing. When they report the strange incident to the police nearby, they are told that it is not the first time.

Japan is rich with ghost stories, especially in Tohoku, which has a long tradition of storytelling. Japanese culture is also thick with superstitions based on animistic beliefs, numerology, language, the supernatural, and death. Akasaka feels that ghosts are a figment of the imagination but serve an important role in the mourning process. "When people are going through a hard time or have lost a loved one, they need a ghostly figure to help them heal their loss," he said. "I have also seen a ghost."[16]

IT WAS THE EVENING OF SUNDAY, MARCH 13, when Setsuko heard the news of Takuya. His body had been found near a river, not far from their home. For Setsuko, it seemed as if he had been trying to find a way back to her and their family.

The next morning, it took all of her strength to go to the elementary school where his body was being held in the makeshift morgue. She was sure it would be mutilated in some way or barely recognizable. *At least his body was found,* she thought. *So many are still missing.* Fortunately she was not alone. Her sister-in-law and a neighbor had agreed to go with her and keep her company.

As she stepped into the gym, the sight astounded her. Corpses, wrapped in large bags, filled the floor. A policeman brought her to Takuya's body. Lying next to him was one of his closest friends. The two had been buddies since childhood. Setsuko marveled at the coincidence. *Even here you're still together,* she thought.

As the policeman carefully opened Takuya's bag, she could see that he was still wearing his work ID tag, which had made it easy to identify him. With fear and hesitation, she then looked into his face but was amazed at what she saw. *He looks so serene, so beautiful,* she thought as she touched his smooth cheeks and closed eyelids. There were no gashes or marks. His neatly brushed black hair was thinning, the only sign of his 58 years. Although his clothes were muddied, his body was not bloated or disfigured. He still looked like the husband she had always known. It was almost as if he was taking a quick nap and would awaken at any moment. Even his watch that he wore every day was still ticking. Tears spilled from her eyes as she felt the weight of despair meld into relief. He had come back to her and the family after his treacherous tsunami journey and was now finally safe, out of harm's way.

On the way to the crematorium on March 20, Setsuko and Takuya's children, son Rigeru and daughter Kotone, held the coffin tight on the back of a small truck as they bounced up and down along damaged and debris-filled roads. They were lucky to have a priest from their Buddhist sect offer funeral rites at the site.

With no cinerary urn available, the threesome improvised with a family flower vase, filling it with selected bones and ashes according to custom.[17] After returning home, they placed the vase in the family Buddhist altar where it would be kept for 49 days. Families living in evacuation centers stored their urns among their few remaining belongings.

Setsuko and Kotone gathered some of Takuya's favorite things to display beside the altar now placed with his framed photo: his well-worn leather briefcase with mobile phone and pen case tucked inside; his work ID; an empty pack of his cigarettes; a can of beer in a holder that he often used during the family's many outdoor barbecues with friends; a penknife because he loved nature, camping, motorbike touring, and gardening around the house; and one of his favorite books, *The Place You Return*, with Setsuko's hope that his spirit would often return home. She wore his watch for a while, thinking it would help her feel his presence, but it depressed her instead. She eventually added it to the rest of their precious memorial collection.

On a clear night, she would look up at the stars and remember the many hours he—a true astronomy buff—spent gazing through his telescope. Even before their son was born, Takuya had chosen the name Rigeru after Rigel, the giant star in the constellation Orion and one of the brightest in the night sky. Kotone he named after the "sound of the koto," or Japanese harp.

Kotone stayed with Setsuko as often as she could, taking the overnight bus rather than the train from her home in Yokohama, near Tokyo, to save on expenses. Setsuko was grateful for her daughter's company. Because Setsuko had lived in Rikuzentakata her whole life, she had many friends and relatives, but everyone was now struggling to survive. She did not want to burden others with her grief. *There are so many worse off than me,* she would remind herself. She kept busy volunteering at the death registry and supporting others. But still, at unexpected moments or when she let herself think about Takuya, the loss and terrible emptiness would well up inside her. She wanted the pain to go away. *Time. It'll just take time,* she thought. *But how long?*

Grieving and bereavement are highly personal experiences, influenced by culture and religion. And yet there are factors that cross these borders, which can play an integral part in the recovery process. "In Tohoku, people feel connected with their community, and that connection will help them overcome their grieving," says Reverend Paul Silverman.[18] An American abbot of a Zen Buddhist temple in Japan, he is now relocated in New York. "Spiritual and societal health comes from integration."

Silverman cites a classic story of Shakyamuni Buddha, approached by a woman grieving over the loss of her child. The great sage told her to collect a sesame seed from every family that had experienced the same tragedy. When she returned to him sometime later, she held a huge vat full of the seeds. Realizing that she was not alone enabled her to heal.

When counseling the grieving, Silverman points to values as the glue. "While you have to be flexible to culture, you need to talk to their values. This will help them in their healing process. In Buddhism, Catholicism, all religions, values and integrity are the same."

Grief counseling, however, is new to Japan. After World War II, talk of death became virtually taboo. The nation had experienced such enormous destruction and the glorification of dying for the emperor and country. The mood turned to revival and the pursuit of economic stability. Sharing private grief over the loss of a relative was also counter to the much-treasured Japanese spirit of *gaman*—endurance. This is especially true in Tohoku where stoicism and self-sacrifice have been especially called for.

Japan has had approximately 30,000 suicides every year since the late 1990s. This is one of the world's highest rates. A government survey found that by June 2011, four months after the Tohoku disaster, the suicide rate in hard-hit Miyagi Prefecture had increased 39 percent. The report warned that stoicism since the disaster may have masked posttraumatic depression. There have been reports of farmers committing suicide after their livestock or crops were affected by radiation from the Fukushima plant.[19] The elderly and those struggling with survivor guilt have also been found vulnerable.

"Hope is the key," says Alfons Deeken, one of Japan's earliest proponents of death education and grief counseling. The gentle, rotund 80-year-old has published 34 books on the topic in Japanese, won numerous awards, and given hundreds of lectures at medical and nursing schools. A Catholic priest and professor emeritus at Sophia University in Tokyo, he first came to Japan in 1959.[20]

Deeken's first encounters with death came during his childhood in Nazi Germany. During World War II, he witnessed classmates burned to death during Allied bombings and watched in horror as his grandfather, an anti-Nazi activist, was fatally shot by an Allied soldier. But it was the death of his four-year-old sister from leukemia, when he was eight, that set him on the path toward his lifework in thanatology (the study of death). His sister's strength in the face of death and her message of hope influenced him deeply.[21]

Deeken began teaching his popular Sophia University course, Philosophy of Death, in 1977, but changing attitudes about death and grieving in Japan has been a slow process. When he established the Japanese Association for Death Education and Grief Counseling[22] in 1983, it was the first of its kind. The year 1986 was a turning point for interest in the issues when he published three volumes on death education that became best sellers.[23] And yet government policy remained passive on the issue.

Compounding this has been the Japanese Buddhist tradition to concentrate more on funerals than on consoling or counseling the living. "Some young Buddhist monks have said to me they'd like to work in hospices, but elder monks told them it wasn't their job," says Deeken. In Shintoism, consoling is limited to the souls of the dead and easing their passage into the next world. Death is treated as impure, and Shinto priests are called on to perform purification rites in places where corpses are found.

From his experience with grief counseling after the 1995 Kobe earthquake, Deeken knew that the Tohoku disaster survivors needed similar support, especially after their sudden and multiple losses. He was surprised at the lack of professional counseling available and blames government bureaucrats. In May 2011, when Deeken offered to counsel Tohoku evacuees relocated to Tokyo, officials told him, "They do not need psycho-emotional support" and his service was "not required." The next morning he read in the newspaper about several suicides in evacuation centers. "The bureaucrats were not even aware of the issues and problems," he says.

Young Japanese are more open minded about death and dying. During Deeken's lectures at various high schools, he has found the students full of questions and willing to engage in discussion. "Young people have a natural curiosity and personal interest in confronting death and dying," he says. He learned that among his 800 students at Sophia University, about 20 percent had experienced the death of a family member, but speaking about it with parents was taboo.

ON MARCH 16, WHEN YUKARI TACHIBANA heard the news that her father's body had finally been found, it was too much to bear. She immediately telephoned her best friend, Chie. The two 18-year-olds had been friends since they were young and thoroughly trusted each other. Yukari knew that Chie would listen carefully and understand her feelings.[24]

Firemen found her father's body stuck on the second-floor stairwell of a fish-processing factory in the town of Onagawa, not far from their home in Ishinomaki. A truck driver for a transport company, he and two colleagues were making a delivery to the factory. The one who survived had urged the other two to come with him to a sports park on higher ground just five minutes away. Yukari's father and the other driver were sure that the two-story factory building was high enough. It seemed far away from the sea and was located on a small hill. The area had not been touched by the 1960 Chile tsunami.

By not heading for higher ground, Yukari's father, like so many, had sealed his fate. In Onagawa, the tsunami had dragged huge fishing boats to the tops of buildings three and four stories high. As the waters receded, there they remained, like oddball Noah's Arks, another of nature's cruel pranks. Yukari tried to block from her mind the horrific image of her father thrashing in the freezing, whirling water as he struggled to catch his final breath before drowning. She often wonders how different things would be if only he had walked five minutes farther up the hill.

Yukari was terrified to see her father's corpse. It took all of her courage to go and identify the body later on March 16 with her grandmother and other family members. The president of her father's company and the man's daughter also joined them. Company staff had found him among unidentified bodies held in a large tent outside a public school. His body was now wrapped in a plastic sheet to which was attached a form listing his age and physical characteristics. When they peeled back the sheet, Yukari and the others stared in shock. He was shirtless. There were no visible cuts or bruises. It seemed as if he were sleeping.

Funeral facilities in Ishinomaki had either been destroyed or were filled with the thousands of residents who had perished in the tsunami. There was one funeral home that could take him in nearby Sendai. They were able to get a time slot at a crematorium on April 5. The crematorium staff was very kind and helped prepare a coffin and urn. When they held the funeral at the family Buddhist temple on July 2, the monks were surprised to see them again so soon. The funeral for Yukari's great-grandmother had been held there just months before the tsunami.

With their family altar destroyed, Yukari and her grandmother improvised with a chest in the tiny living room of their temporary housing. Along with her father's photo set in a thick black frame, it holds an impressive collection of memorial tablets of her many ancestors. Her grandmother is the keeper of these family tablets. They will be passed on to Yukari, the only direct descendent of a one-hundred-year-old lineage of kimono makers. The family had certainly seen better days.

Yukari was raised by her grandmother and great-grandmother after her parents divorced when she was three. Her father had been in and out of her life, but she still felt close to him. Her mother disappeared from her life after the divorce, and her father remarried and divorced again. Yukari cannot recall her mother's face and has no photo to remember her by. She is not even sure if her mother is still alive. With both parents gone, she could be considered an orphan.

"I don't feel like an orphan," she says firmly. She still has her grand-mother and other family members close by. "But I know how very difficult it would be if a child became an orphan after losing both parents in the tsunami."

IN RIKUZENTAKATA, AFTER A LENGTHY SEARCH through the town's 130 evacuation centers, remaining homes, and recovered family re-cords, volunteers from child support centers found 41 children under the age of 18 who lost both parents. Toshinao Kanno, head of the city's welfare bureau, was faced with a daunting task. There was no financial support set up for this kind of situation. "Unlike big cities like Tokyo, though, we have a very close community. It's unthink-able that a child would be neglected or not taken in by relatives," he says.[25]

Most of the orphaned children in Rikuzentakata were able to move in with relatives, despite cramped conditions and financial con-straints. The remaining few, like 11-year-old Yuki Sato,[26] were sent to foster care institutions. Since May 2011, she has been living at the La Salle Home in Sendai, part of a worldwide, Catholic-based congregation of schools and foster homes. "She's doing beautifully," says Brother Rodrigo Treviño,[27] director of the school and 20-year resident of Japan, originally from Mexico.

Like many children, Yuki was at school when the tsunami hit, just a few days before her fourth grade finished. Along with her par-ents, she lost her four-year-old sister and grandparents. She stayed briefly with two uncles and their families in Sendai, but both men lost their jobs after the disaster and were pressed financially. With La Salle Home close by, she is able to see them and her aunts and cousins often.

Treviño admits that there are few options available to orphans in Japan.[28] "Japanese nationals normally do not adopt. It's not part of their culture. Foreigners adopting Japanese children is almost nonexistent," he explains. La Salle Home can take care of Yuki until she is 18, and then by law she has to leave. The adjustment can be traumatic. The home tries to ease the transition with a one-year "after care" program, and Treviño says its doors are always open to all of its "graduates" as a temporary place to stay. "We recently had some stay with us after they lost their job," he says.[29]

The home has about 80 children—the maximum allowed. Yuki is the only orphan who has lost both parents.[30] Treviño says they are closely monitoring her condition while also advising her uncles and coordinating with the local welfare bureau. Fortunately she is not being bullied at school, an old problem for kids from foster institutions. She meets with a therapist once a week. A counselor will sometimes join her for a meal or when she is studying. She even has a sponsor. The athletic equipment company ASICS has donated her sports gear.

Yuki is still assimilating the impact of the disaster. The staff at La Salle Home have not seen her suddenly cry or heard about any nightmares. She quietly revealed that she felt better after her father's body was found. Her counselors say that to some she can appear cold because she does not talk much about her family. As a child, she is brave and strong. During her teenage years, she will struggle more with her loss.

"ADULTS TEND TO MINIMIZE CHILDREN'S CAPACITY to understand and deal with tragedy," says Kiku Iwamoto of the Tohoku relief team of NGO Peace Winds Japan. Kiku had come to the hard-hit region to set up a soft, post-traumatic support program for children and young teens introduced by Portland, Oregon–based non-profit Mercy Corps. Called "Comfort for Kids," the program was developed by Mercy Corps after the September 11, 2001, terrorist attacks to train caregivers and parents in ways to support children affected by trauma.[31]

Kiku recalls a powerful conversation she had with several children who calmly told her about their family members and homes that had been swept away by the tsunami. "They didn't show their sadness by crying like adults," she says. "It takes children time to process it all." Later, Kiku spoke to the mother of one of the boys. The mother had not talked to her son about the tragedy because she was afraid he would get upset. "The boy didn't have a chance to explain his feelings to his mother because she never asked him."

To set up the program, Mao Sato of Peace Winds Japan first approached the local boards of education.[32] She was told that the national government would be offering a psychosocial program, and it would be difficult for an NGO to do the same. "But the government program wasn't implemented until April 2012, one year later, because of the overwhelming number of people affected by the tsunami," says Mao.

Japan does not have a history of well-developed psychosocial support programs for children. Public schools are normally staffed with a guidance counselor, but teachers are often asked to help troubled children and their families. In the weeks after the disasters, there were few who could fill the role of counselor or caregiver. The schools were also busy trying to reopen a month later. Emotional care was low on their list of priorities.

"Educators in Japan are very concerned about children's mental health, but they didn't recognize the advantages of working with NGOs," says Mao. It wasn't just Mercy Corps. UNICEF and Save the Children were also faced with similar difficulties in working with the prefectural and national governments. Peace Winds Japan decided to bypass the school system and work with the evacuation centers instead. Trained volunteers from Tokyo and other locations outside the hard-hit areas were brought in to interact with the kids. Once the centers were closed in August 2011, they set up at temporary housing units.

Three programs were created. One for sports, called Moving Forward, was in cooperation with the local Junior Sports Club Association and sponsored by Nike. Another, called Art Caravan, was a portable, child-friendly space with art supplies where children could express their feelings through drawing and painting. The third program was Ochakko ("teatime" in the Tohoku dialect), which was a program for adults that provided a spot near the Art Caravan for parents to sit, chat, watch their children, and enjoy a cup of tea. This way, counselors and instructors were able to establish a relationship with the children and parents directly.

There were children who clearly benefited from the programs. Mao recalls one sixth-grade girl who lost her father in the tsunami. She had been the captain of a sports team and joined the Moving Forward program. "She was so strong. Didn't cry and complain, but she seemed tense and didn't talk with anyone about her loss," says Mao. "We had some young and very friendly local instructors who she opened up to. She started smiling and laughing and even helping the other kids. We could really see the difference."

The children had to make major adjustments when they moved out of the evacuation centers to temporary housing. "For many children, life in the evacuation centers sheltered them from what was really going on because they were surrounded by friends from school all the time," says Kiku. "It was sort of like camping," a boy told her. The temporary housing was chosen by lottery, so children were usually separated from friends and their former community. They were not used to dealing with strangers. "When the kids moved to the temporary housing, reality finally hit them," she says. At first many kids withdrew and didn't play outside. Space was limited, and they were scolded if they made noise. "But little by little they adjusted to their new life," she says. "They started establishing a new routine and getting control of their lives and schedules. They also found someone new they could play with."[33]

Kiku saw amazing resilience. "A lot of people were looking for posttraumatic stress disorder. But what I saw was resilience and how strong people were, including little children," she says. "Instead of PTSD, I believe in 'post dramatic growth idea.' It's an idea I learned from the Dougy Center, that people have power to face their own grief and gain control of their lives."

AFTER SCHOOL RESUMED IN APRIL, David Chumreonlert could not detect posttraumatic stress in any of his students, mainly because he did not have a chance to spend time with them individually. He did sense a hesitation to talk about the disasters, however. "A few students casually said, 'My family used to be three and now we're two,'" he says. He felt awkward about probing further.

Toru Saito's maternal grandfather was swept away by the tsunami. It was probably a helicopter or patrol boat that found the body floating on the sea near the Sendai port in late April. Toru's family was contacted on May 15 after the teeth had been identified from X-rays. They exhumed the deteriorated body from a temporary graveyard in Onagawa and had it cremated on May 21. The funeral was held on May 22. Toru does not mind talking about his grandfather's death or the disaster. He and his friends, he says, have moved on. But his pensive gaze as he walks slowly across the gravelly dirt lot that was once his home reveals the weight of his loss. Time may heal the trauma, but Toru knows he must never forget.

TEN

Tohoku Damashii

There is nothing to be frightened of. Compared to the great virtue that envelops the world, your sins are what a little drop of dew on the point of a thistle's thorn is to the light of the sun.

—*Kenji Miyazawa, "Bare Feet of Light"*[1]

Relative Levels of Radio-activity in Disaster Zone

Three-dimentional visualization of monitored data in April 2011.
Note that the height of the maximum level at the Fukushima Daiichi Nuclear Power Plant exceeds the height of this page by a factor of ten.

•Rikuzentakata

•Sendai
•Oginohama

•Soma

•Minamisoma

•Okuma •Fukushima Daiichi Nuclear Power Plant

•Iwaki

ON THE MORNING OF MARCH 11, 2011, MINAMISOMA and its mayor were grappling with the same mundane problems as other small rural cities across Japan: a declining, graying population, creaking public services, and a faltering local economy. By nightfall an existential disaster had engulfed Mayor Sakurai's office.

The earthquake and tsunami took 947 lives, including about 100 children, in Minamisoma. Fallout from the Fukushima Daiichi plant contaminated the city and sent most of its population fleeing. As of April 2012, 27,000 residents—a third of the population—have scattered across Japan and as far away as North America and Europe. Hospitals and schools have installed dosimeters outside, showing airborne radiation in blinking red figures. About 150 of the city's 830 employees have quit in what the mayor calls a "municipal meltdown" brought on by the enormous stress of the 2011 calamity.

Minamisoma's agony is replicated along the northeast coast, where 19,000 people are dead or missing.[2] The deluge has left behind gaping landscapes reminiscent of the atomic aftermath in Hiroshima and Nagasaki. Journalists found car navigation systems in many coastal towns and cities still directing them to landmarks that had vanished. For months, survivors picked through the mud for belongings. In makeshift refugee centers, photographs plucked from the sludge were laid out at the entrances in the hope that their owners would claim them, if they had survived. Across the region, family photos were placed on piles of debris, along with snacks and fruit, impromptu altars to hundreds of anonymous loved ones. On the coast, families could often be seen tossing flowers into the sea as they struggled to find a focus for their grief.

That detritus of memory has been cleared away. Only the rusting steel spines of the strongest buildings now stand in many of these coastal towns in Iwate, Miyagi, and Fukushima Prefectures. Temporary, two-roomed homes have sprung up in schools, parks, and every available public space, housing the roughly 340,000 people displaced by the disaster. As we write, 22 million tons of rubble is piled up on the outskirts of most coastal towns and cities. Local governments around Japan had refused to remove any of it because of fears of radioactive contamination, though they have slowly started to relent.

Soma was one of 300 fishing ports damaged or destroyed, along with 22,000 boats. Thousands of fishermen along the Tohoku coast have confined themselves to harbor because of radiation fears. The total blow to Tohoku's fisheries from the tsunami in a country, and region, synonymous with the sea and its resources, is estimated at ¥1.2 trillion—nearly $15 billion.[3] The long-term impact is much harder to predict. Fewer than 2 percent of Fukushima's fishermen have returned to the sea, and, even if they have the boats and equipment to do so, they are unable to sell their catches.[4]

Perhaps a third of the remaining fishermen in Tohoku will quit, driven from the industry by the disaster.[5] For men such as Yoshio Ichida, without sons to take over, the motivation to go into debt and rebuild their businesses is not strong. "Even if you fish, you don't know if it will sell," he says, echoing the plight of his colleagues. "As long as the reading shows any level of radiation, I wouldn't let my grandchildren eat it." Eventually, if leaks from the nuclear plant and its roughly 200,000 tons of contaminated onsite water cease, seaborne radiation will fall. In June 2012, commercial fishing finally resumed, but it was limited to octopus and sea snail, two species considered less vulnerable to radiation. The reputation of local seafood, however, will surely take much longer to recover.

History has shown that Tohoku communities can rebuild, often with remarkable speed. Recovery this time, however, is less easy to predict. Even before March 11, 2011, communities such as Minamisoma and Rikuzentakata, where more than a third of the population is 65 or older, were withering. The disaster is likely to accelerate migration to the cities, further sapping the region of the energy and taxes needed to rebuild. Most towns and cities will never be the same. Some may cease to exist altogether.

The disaster struck at an ominous, possibly defining moment in Japanese history. At the outset of its journey toward modernization in the 1860s, Japan's population was 30 million. Today, after the end of its remarkable postwar boom, over 127 million people live on the crowded Japanese islands, straining land and resources to the limit. Populations have moved closer to the vulnerable coast since the great tsunami disaster of 1896. Japan's network of nuclear plants was driven by dangerous expediency: the need to power the country's long postwar economic boom in a country with little oil and a shortage of suitable sites. By the turn of the twentieth century, the boom had long peaked, along with the population. In 2012, the Ministry of Health, Labor, and Welfare predicted that the number of Japanese will plummet by 30 percent in the next half century, while rising life expectancy further burdens the state. The shrinking and aging population will leave the government grappling with ballooning social welfare costs while trying to pay off Japan's public debt—the highest in the industrialized world. Even had the tectonic plates not shifted in March 2011, Japan would be facing hard choices.

Recovery is complicated by the radiation from the Daiichi plant, which has blanketed 8 percent of the entire country. The impact of the radiation is bitterly disputed and is almost certainly less serious than feared, but the insidious and excruciating uncertainties are in themselves hugely stressful for those living with the fallout. As Hiroaki Koide, nuclear reactor specialist and assistant professor at the Kyoto University Research Reactor Institute, said: "This is something which humanity as a whole has literally never experienced, we have no idea—I have no idea what will happen."[6] The first large-scale government survey of Fukushima residents, in the heavily contaminated Iitate Village, Namie Town, and surrounding areas, found cesium–137, which has a reported half-life of about 30 years, in urine samples from 32 out of 109 people.[7] Cesium–134, with a half-life of about two years, was found in nearly half the test subjects. Screenings of children for iodine–131 in Iitate and 15 other municipalities from March 26 to 30, 2011, found exposure of up to 0.1 microsieverts, "equivalent to an annual dose of 50 millisieverts for a one-year-old."[8] Children, who are closer to the ground and therefore ingest more toxins, are especially vulnerable to the effects of radiation. Small quantities of radioactivity were detected in breast milk in the months after the disaster, forcing the government to announce publicly funded tests.[9]

The government's March 2011 evacuation directive has emptied once-thriving towns around the plant, like Kai Watanabe's hometown of Okuma, of 114,000 people. An unknown additional number—anywhere from 50,000 to 120,000—have moved voluntarily because of radiation fears, ignoring official claims that life inside or around Fukushima Prefecture is safe.[10] Mothers who stay must warn their children of the dangers while trying not to make them, as one commentator noted, neurotic about every particle of air, soil, water, or food around them.[11] Thousands have taken their children and started new lives elsewhere, splitting up families, often against the wishes of protesting in-laws and fathers, who are tethered to work or simply less concerned about the nuclear accident. The incidence of suicides has risen, especially among the old and infirm, ripped out of local communities and living in temporary housing. In one particularly horrifying episode, a depressed middle-aged woman set herself on fire in one of these cramped houses; her family is blaming, and suing, TEPCO.

Masked workers in white boiler suits initially descended on schools, parks, and government buildings with power hoses and mechanical diggers, scouring and scraping away the contaminated soil. Across Tohoku, trucks loaded with the soil trundle to radioactive dumps inside the exclusion zone. The work has helped create a construction boom in Tohoku and will leave a pile of nuclear waste large enough to fill one of Tokyo's largest stadiums 80 times, according to the local media.[12] There is widespread doubt about how successful the strategy will be. Government regulations stipulate that about 2 inches of soil must be scraped away, but many farmers say the contamination has sunk much farther into the ground. While the forests of Iitate Village and other mountainous areas wait for the workers to come, the toxins will wash down into the land, contaminating it once again.

Inevitably, perhaps, in a country with a unique nuclear legacy, Japanese popular culture is full of eerie warnings about such a disaster, from the radiation-breathing Godzilla to the Hayao Miyazaki animation classic *Nausicaa of the Valley of the Wind,* set in a postapocalyptic world where settlers live in constant fear of the encroaching poison from the toxic jungle. The plant's crippled, unstable reactors recall the giant caterpillarlike Ohmu insects in *Nausicaa,* ready to rage and cause deadly damage at the slightest provocation. The message is usually the same: Humans pay the price for disrespecting nature. In the Kenji Miyazawa story "The Bears of Mt. Nametoko," the hunter Kojuro kills bears for their livers and is eventually killed by the bears that were his victims. Above his corpse, the heavens seem to be passing judgment: "The Pleiades and Orion's belt twinkled green and bitter orange, seemingly breathing in and out."

Mayor Sakurai, like many local leaders, is squeezed between the pressure to resuscitate his city and to protect its remaining citizens. The city office says the most dangerous contaminant, cesium, has been removed from the grounds of Minamisoma's Middle High School. A large dosimeter is planted outside the school's front gates, showing the radiation in big digital letters, to reassure local parents—with limited success. In March 2012, a year after the disaster, only half of its 360 students had returned. The flashing red dosimeter reads 0.2 microsieverts, lower than the background radiation in many Western cities.

In March and April 2011, however, the radiation was ten times that, exceeding central government guidelines of 1 millisievert of radiation per year. In response, the government hiked the maximum allowable limit to 20 millisieverts, sparking bitter protests. Thousands of people across Fukushima voluntarily evacuated as far as Kyoto, Kyushu, even Okinawa, 1,200 miles away. Parents staged weekly demonstrations outside the education ministry's offices in Tokyo, where the guidelines were set. Mothers who had lived quiet, rural, middle-class lives before March 11, 2011, found themselves in the capital, screaming invectives at elite bureaucrats. Children who remained in Minamisoma were limited to playing outside for two hours a day until April 2012, when the mayor partially loosened restrictions. Some parents, armed with their own dosimeters, say radiation is still too high to allow their children to attend school. The mayor, who is unmarried and childless, is underestimating the dangers, they say.[13]

A few miles from Minamisoma's high school, National Route 6, which runs right by its gates and down the Pacific coast, hits the 12-mile exclusion zone around the crippled plant. The highway can be driven almost to the plant's hulking corpse before it collapses into quake-induced rubble, but police officers with dosimeters pinned to their chests prevent locals from entering, even Mayor Sakurai, whose farm is just on the other side of the evacuation zone. Reporters and citizens who venture inside through back roads can be arrested. Inside the zone, life has frozen in time. Homes have been abandoned and reclaimed by weeds. The 80,000 people who once lived here have not been told when, if ever, they can return. In a desperate attempt to keep their communities alive, the mayors of Okuma, Namie, Futaba, and Tomioka plan to create "temporary" towns elsewhere.

Tohoku was one of Japan's last frontiers. The region has a rich, vibrant culture but a history fraught with poverty and struggle. This and the harsh environment have shaped the resilience of local people. Their strength is such that the Japanese refer to it as *Tohoku Damashii*, the "spirit of Tohoku." Master haiku poet Basho Matsuo (1644–1694) was one of the first to note the spirit of the region in his famous journal *Oku no Hosomichi* (The Narrow Road to the Deep North), about his journey through the "back country" of Japan.

Writer, passionate Buddhist, and social activist Kenji Miyazawa (1896–1933) is sometimes seen as the embodiment of the Tohoku spirit and is the region's de facto poet laureate. He devoted the final years of his short life to bettering the lives of poverty-stricken peasants in his native Iwate Prefecture. His writings, oddly prescient of the disaster that unfolded nearly 80 years after his death, have become a touchstone for its beleaguered natives since March 11. In *Night on the Milky Way Train,* a story about the death of a child and overcoming grief, Kenji writes: "I'm not scared of that dark. I'm going to get to the bottom of everything and find out what will make people happy. Let's go together . . . as far as we can go." "Strong in the Rain" became the signature poem of the disaster, its spare words seen both as a peon to the resilience of Tohoku people and an invocation to further struggle. Some locals have grown to dislike the poem because its message of self-sacrifice and individual endeavor, quoted

so relentlessly in the national media, can seem like another directive
to *ganbaru*—to keep going, endure, never give up—when many feel
like doing precisely that.

A millennium ago, the indigenous Tohoku *Emishi* resisted imperial rule but were conquered by the reigning Yamato state. From
the fifteenth to the seventeenth centuries, the region was drained
by debilitating taxes used to fill war coffers. The Boshin Civil War
(1868–1869) marked Japan's entry into the modern era. During this
time, regional lords picked the wrong side, aligning with the weak-
ening shogunate rather than ascendant imperial forces in Tokyo. As
Tohoku specialist Norio Akasaka notes, the region has always been
on the losing side of history.[14]

The severity of Tohoku's geography and climate has also honed
the local instinct for survival. The region's winters are harsh and
long. Cold winds blowing down from Siberia mix with moisture
from the Japan Sea and bring the largest snowfall in the world to
the northwestern coastland and mountains. Three parallel mountain
chains divide the region, with the Pacific Ocean bordering the eastern coast. Cold sea currents have created one of the world's richest
sources of seafood.

But it is the region's vast rice farming that has weighed heaviest on Tohoku history. Lauded as one of the country's breadbaskets,
as early as the 1600s, edicts from ruling shogun and regional lords
pushed local farmers to produce enough rice for growing demand in
the capital of Edo, now Tokyo. Fickle weather and incessant taxation
also produced periods of terrible famine.

This service to Tokyo continues today. Before March 11, Tohoku provided a quarter of Japan's rice crop. Government subsidies
backed feeding the capital. Towns and cities along the northeast
coast had famously thriving fishing ports. Fisherman Ichida's floun-
ders and sardines were sent to Tokyo's Tsukiji, the world's biggest

fish market. Fukushima's reactors too were built specifically to service the energy-hungry capital. The Daiichi plant produced not a single watt of electricity for its host, though it brought billions of yen in local subsidies—a Faustian bargain that many locals now bitterly regret. In a bitter irony, some farmers and fishermen who have been put out of business by the radiation have been tempted into working at the Daiichi plant. Flyers posted around parts of Tohoku advertise the work for ¥14,000 yen (about $175) a day. "Middle-aged men especially welcome," say the flyers, because they are less vulnerable to the effects of radiation than younger men.

An old saying among the people of Tohoku self-mockingly described their historical role to serve Tokyo with soldiers from their men, prostitutes from their women, and rice from their farmers. Add nuclear energy and Tohoku's cheap labor force emigrating to the capital, and it seems to many locals that Tokyo still relies heavily on the tough "country folk" in the northeast, even though the region today accounts for only 4 percent of Japan's gross national product.

The sense among some Tohoku citizens that they are again prey to the arrogance of the imperious capital south is reinforced by what they see as the unsatisfactory nuclear compensation paid out by the Tokyo-based TEPCO. By summer 2012, most victims had received less than $20,000 (¥1.6 million). Many had to wait until September 12, half a year after the accident began, for the claims process to properly begin. The utility started sending, mostly through the mail, 58-page application forms for compensation that demanded receipts for transportation and other fees incurred during the evacuation, bank or tax statements proving pre-disaster income levels, and documented evidence of worsening health since the move.[15] A month later, TECPO had received just 7,600 completed forms—about 10 percent of those sent out—because they were widely considered too arduous and detailed.[16] The explanation book accompanying the application form came to 160 pages. Amid withering criticism, TEPCO explained that it was merely trying to cover all bases, but it was forced to radically simplify the application procedure.

TEPCO's compensation scheme cleaves closely to the government directive on evacuation, meaning that only those who have been *compulsorily* moved are entitled to claim damages. The utility sidestepped the question of abandoned property and other assets, since the government line is that evacuees from Okuma, Futaba, Iitate, and other heavily contaminated areas will someday return to their homes, farms, and ports—something few scientists believe is either possible or desirable. In July 2012, the government finally accepted that areas contaminated with annual exposure of over 50 millisieverts will be uninhabitable for at least five years. A zoning system divided the fallout zone into three areas on a sliding scale of contamination. The move meant that Kai's family can finally claim compensation for their Futaba home.

But what about Iwaki and Minamisoma, whose mayor announced that he is suing TEPCO for economic damages? About 27,000 of the city's population of 70,000 may leave permanently, depriving it of taxes and perhaps bringing bankruptcy.[17] Who will pay for that?

The compensation scheme takes no account of the long-term impact of prolonged exposure to low-level radiation.[18] Eventually, these claims will begin, and bureaucrats and lawyers will grapple with the maddening, inexact science of judging the impact of cesium, strontium, and plutonium on human populations. Japan has a 20-year limit from the date of an accident for compensation claims. No one knows what will happen after that time.

Estimates of the total cost of the Fukushima catastrophe, including compensation, fluctuate wildly. TEPCO was told by an advisory panel in October 2011 to prepare for claims of ¥4.5 trillion yen ($56

billion) in the two years until March 2013. The broadest calculation puts the entire cost of the disaster, including compensation and
decommissioning the Daiichi plant's six reactors, at ¥40 to 50 trillion (US$504 to $630 billion)—a figure that approaches the bill for
cleaning up the US subprime banking meltdown in 2008 and 2009.[19]

Who will pay for it? TEPCO has already argued in court that it
is not responsible for the radioactivity showered across Fukushima
because it doesn't "own" it. "Radioactive materials . . . that scattered and fell from the Fukushima No. 1 nuclear plant belong to
individual landowners there, not TEPCO," the utility's lawyers told
Tokyo District Court during a disposition to hear demands by the
operators of the Sunfield Nihonmatsu Golf Club, 28 miles west of
the plant, that TEPCO decontaminate the property. The owners said
they were "flabbergasted" by the argument, but the court essentially
freed the utility from responsibility.[20] If the decision holds through
legal challenges, local and central governments will be forced to foot
the bill. In May 2012, to nobody's great surprise, Japan's government
began nationalizing TEPCO, injecting ¥1 trillion ($12.5 billion) into
the essentially bankrupt utility, one of the largest bailouts in recent
history.[21]

The victims of the Fukushima nuclear disaster face a choice of
either waiting, if they are entitled under the guidelines, for a TEPCO
settlement to their claims or going to court. Some will fight. The
strategy of TEPCO and the government during what is likely to be
the most expensive liability case in Japanese history is, in effect, to
suppress compensation claims by making them as restricted, bureaucratic, and difficult as possible for thousands of Fukushima victims.
In the most famous Japanese mass compensation case of all—the
mercury poisoning of food around the town of Minamata in the
southern island of Kyushu in the 1950s, it took over 40 years to
settle claims. Even today, new claimants emerge. Most victims of the

Fukushima accident, such as Kai's parents, have for now abandoned hope of going back and gotten on with their lives. As their family restaurant slowly disintegrates in Okuma, they have started working as school caterers in Iwaki City.

The vast, unspooling legacy of Fukushima is particularly poignant in a country with a unique experience of man-made nuclear catastrophe. Hiroshima and Nagasaki killed or injured at least 225,000 people and left behind 430,000 *hibakusha*, the name Japan gives to atomic bomb survivors. The catastrophe ignited a new science to monitor the impact of radiation on victims over years and decades and, inevitably, politicized this impact between pro- and antinuclear scientists. Since March 11, 2011, many observers have begun referring to another generation of Fukushima *hibakusha*, who must also endure a lifetime of worry, state support, and even discrimination.

Can Tohoku recover? Inevitably, the question leads to an even more fundamental one: What sort of country does Japan want to be? The nation's epic industrialization drive seems to have run out of steam. Its dream of energy self-sufficiency lies in ruins. Its population is aging and declining. Japan's squabbling political leadership seems powerless to stop the nation's slide down the economic league tables. By 2050, Japan may no longer even be considered a developed nation.[22] The stench of national decay seems to run all the way from the still-decimated fishing ports of Tohoku into the country's political and economic heart a few hundred miles away.

The seeds of a radical departure from the past can be seen in Tohoku. In the aftermath of the March 11 disaster, Fukushima's local assembly adopted a petition calling for the scrapping of all ten reactors in the prefecture, the first political salvo in a nationwide movement against nuclear power. By March 2012, one year after the disaster, all 54 commercial reactors were offline, leaving Japan nuclear free for the first time since 1970. But local governments across Japan have no legal tools to veto the restart of the reactors, only moral and political power. One of the last acts of Naoto Kan's government in the summer of 2011 was to pass legislation forcing the nation's utilities to buy more renewable energy. Huge new solar power stations are being built in northern Japan and elsewhere. Remarkably, in June 2012, Toshiba, which installed two of the Daiichi plant's reactors, announced it is building Japan's largest solar power project on Minamisoma's ruined coast.

The power centers in the capital to the south have signaled their displeasure. Exactly a year after the Fukushima crisis erupted, Prime Minister Yoshihiko Noda, who replaced Naoto Kan, said that "no individual was to blame" for what happened, essentially absolving the nuclear establishment of responsibility and helping to ensure that none of its leaders would be prosecuted.[23] By 2012, nearly half the 20 executives who quit after the Fukushima crisis had found lucrative positions elsewhere. Disgraced former President Masataka Shimizu was appointed an outside board member of Fuji Oil Co., a company with which TEPCO has a close relationship. Fuji Oil said the company wanted to use Shimizu's "profound experience in the energy sector."

The central government, backed by Japan's biggest business lobby and the country's highest circulation newspaper, *Yomiuri Shimbun,* has announced that the reactors must restart. Ending the nuclear dream would mean scrapping billions of dollars in capital investment, withdrawing from an industry in which Japan is now a world leader, and damaging national competitiveness. The crisis has already torn a hole through Japan's commitment to cut greenhouse gases and increased its bill for oil and gas imports by $100 million a day, leaving the country with its first trade deficit in three decades. And, as of July 2012, fuel imports had pushed the trade deficit to nearly ¥3 trillion (US$38 billion), a record amount.[24]

Still, Mayor Sakurai, like many in Tohoku, now believes the nation must steer an alternative path. He wants to turn his city into a global center for renewables and believes the crisis will galvanize a nation that has seemed adrift for two decades. Japan leads the world in the technologies for solar and geothermal energy. Thousands of young people have volunteered for work in Tohoku; university students have begun abandoning nuclear engineering for other careers; farmers are experimenting with alternative crops; schoolchildren have begun thinking about new approaches to the country's problems. "Those are the energies we have to draw on," the mayor says. "That's our future."

The tsunami-ravaged coastline of Tohoku also faces a critical choice: rebuild towns, cities, and villages behind even bigger concrete walls, or move back from the sea and find a way to live with nature rather than going to war with it. Much of the worst-affected areas were natural lagoons in the nineteenth century, reclaimed for towns and rice paddies during the great expansion of the twentieth. The coastal areas should be returned to the sea now that the population is contracting, insists Tohoku native and historian Norio Akasaka. There's not enough concrete to build tsunami walls high enough, even if the region could withstand the destruction of its breathtaking Pacific scenery.[25]

Citizens in mountainous areas outside Fukushima City have won funding from the Environment Ministry to turn a hot springs resort into a center for geothermal power generation, targeting total generation of 1,100 kilowatts per hour, enough to cover all the electricity used in the area. Iitate Village farmer Nobuyoshi Ito, who refused government directives to evacuate, wants to plant thousands of acres of sunflowers, which have been found to absorb radiation.[26]

Tohoku environmentalist Shigeatsu Hatakeyama advises paying more attention to how properly maintained mountain forests can shelter homes from the elements and replenish the seas with nutrients. Chosen by the United Nations as the recipient of one of its "Forest Heroes" awards, which celebrates those who dedicate their lives to replenishing forests, he says building higher walls and floodgates will separate the mountains from the sea and destroy marine resources. "To us, it seems far more important to think about how to get along well with the sea and the mountains than consider how to deal with the threat of tsunami."[27]

Such thinking resonates with Tohoku fishermen like Ichida, who has spent the year since the sea took his boat, home, and friends wondering what the future might bring. "Instead of lamenting that the resources are drying up, I think from now on, we should think about how to properly use the resources, especially after they have recovered from the fishing ban. Previously, we've been going out to the sea to get money. It was a matter of demand and supply; if you fish more, the price of fish declines. From now on, we have to draw a strict line in order for everyone to be able to fill our stomachs rather than recklessly trying to fish as much as possible like we used to."

The ghost of Kenji Miyazawa seems to lurk behind such sentiments, articulating a philosophy that may have been 80 years ahead of its time. A deeply spiritual man who was also a rationalist, scientist, and nature lover, Kenji believed that the Japanese would "die in spirit" if they did not learn to coexist with nature. Once they remembered that old lesson, however dark their suffering may be, "it is only a speck in a vast universe of healing and light."[28]

Stop Working

Stop working
Throw down your rake
I had all the fertilizer planned
Feeling responsible for the rice plants

And the paddies were flattened one after the other

Thanks to this morning's violent thunderstorm

And this half-moon cloudy sky

The plants came down one after another

It is not only in the factories

That work can be demeaning

What is ignoble is

To strive to conceal your insecurity

Working yourself to the bone

Now, get yourself on home

Phone the weather station

Bundle your head up tight

Prepare to be soaked to the quick

Get yourself out and confront each and every person

All the many faces that are stiffened and wan

Go around encouraging them with your fire

Tell them that you will provide them compensation

Whatever it takes out of you.[29]

Epilogue

AS ALWAYS, MAYOR KATSUNOBU SAKURAI CAME OUT fighting—and running. Although his city had gained unwanted global fame as the emblematic victim of the twenty-first century's first major nuclear disaster, he remains determined that it will bounce back. Instead of fleeing from the toxic air, he pounded the city's back roads throughout the winter after the quake, training for his umpteenth marathon. Before heading south with a team of local people for the Tokyo Marathon in late February, the 56-year-old had red sashes made by his city because he knew the nation's television cameras would be in the capital. The sashes read, "With one heart, for the revival of Minamisoma."

For some Tokyo citizens, it was the first time they had seen the mayor without a dosimeter hanging from his neck. His finishing time of four hours and nine minutes put him two hours behind the marathon leaders, but the real performance came in the press conference afterward. *TIME* magazine had helped make him an icon of the 2011 disaster by putting him on its list of the world's 100 most influential people, so in Japan he was now something of a celebrity. Reporters peppered him with questions after the race. Yes, he admitted, Minamisoma's plight was still dire, but there was plenty of cause for optimism. "I'll never give up trying to save my city," he said.

The mayor's optimism was as inspiring as his problems are formidable. Decontaminating the city could take up to four decades, said experts; the population figures might never recover. Children would have to stay indoors for years to avoid hot spots. Who would pay for reconstruction? Houses are being rebuilt, replied the mayor. Medical screenings of all citizens would ensure the safety of local people. Schools had been decontaminated. New plans were afoot for renewable-energy businesses. Farmers and fishermen would eventually begin work again.

In addition to dealing with the disaster that struck on March 11, Mayor Sakurai is struggling with the aftermath of his previous life. The court that had ended his fight against the industrial waste plant has ordered him to pay costs, which would have to come out of his own salary since the suit was in his name. His farm is still unworkable, and he has not been compensated. His aging parents live in cramped temporary housing in the city. And what of the Daiichi nuclear power plant 12 miles down the coast?

The official line since December 2011 was that the plant had achieved a "state of cold shutdown," meaning that radiation releases are under control and the temperature of its nuclear fuel is consistently below boiling point. But the term is considered controversial. Engineers have only a rough idea of where exactly the melted fuel lies inside the damaged reactors and of its exact state. The fuel is being kept cool by thousands of gallons of water that TEPCO pumps onto it every day and which it is struggling to decontaminate. About 200,000 tons of toxic water have built up on-site, stored in huge 1,000-ton water tanks that have been packed into all the available space at the Daiichi site. Nobody knows where the water will go.

Japan's government says that dismantling the reactors and its 260-ton payload of nuclear fuel will take up to 40 years, even if there are no roadblocks in what is an enormously complicated technical task. The final official tally for escaped radiation from the plant is 900,000 terabecquerels, about one-fifth the amount released by Chernobyl.[1] The impact of this radiation on human health will be bitterly contested, as it is 25 years after Chernobyl. About two million people in Ukraine are still under permanent medical monitoring. Among children, monitoring is recommended for about 400,000 who are believed to have received substantial levels of radiation to their thyroid. In neighboring Belarus, cases of thyroid cancer increased after 1986. Estimates of eventual fatalities range from dozens to hundreds of thousands.[2]

There is renewed concern about the crippled building of reactor four, where about 1,500 fuel rods are stored in a spent fuel pool. About five times that number are stored in a common spent fuel pool about 165 feet from the reactor, which is enough to poison not just Japan but much of the world beyond its shores. The damage from last year's disaster left the rods without a protective containment vessel. Another strong quake could cause a "global catastrophe like we have never before experienced," warned Japan's former ambassador to Switzerland, Mitsuhei Murata, at a public hearing in March 2012.[3]

Whatever happens, many people now believe that the government and TEPCO will eventually be forced to recognize that people like Kai Watanabe, who fled from radioactivity a year ago, may not return for decades. The ripples from the disaster continue to spread around the world. Germany, the world's fourth-largest industrial nation, committed in June 2011 to replacing nuclear power with renewables by 2022. Italy voted overwhelmingly by referendum the same month to abandon plans to restart its nuclear program, and Switzerland ordered a freeze on the building of new plants. Thailand and China put nuclear plans on hold, almost certainly temporarily. But remarkably, Japan has begun restarting its idling reactors as of May 2012.[4] The first to start was in the picture-postcard town of Oi, a small fishing town sheltered in a rugged cove ringed by rice paddies and mountains over on the other side of the country from Fukushima. The government order to restart the reactors in June triggered Japan's biggest protests in decades. Throughout the summer young and old gathered outside the prime minister's office in central Tokyo to shout antinuclear slogans, to widespread media indifference. An unprecedented antinuclear rally of perhaps 170,000 people in July 2012 was ignored by *Yomiuri Shimbun,* Japan's highest circulation newspaper. Prime Minister Noda finally agreed to meet some of the protestors in mid-August. It was a rare Japanese accommodation to people power but Noda was vague, saying only that the government was "considering" energy policy "with a view" to phasing out reactors. One reason for the stubborn political support for atomic power is of course the cost of pulling out. Another is defense. "Having nuclear plants shows that Japan can make nuclear arms," said former defense minister Shigeru Ishiba the same month. "Japan's plutonium stockpile 'works diplomatically as a nuclear deterrent,'" admitted the *Yomiuri* in 2011.[5]

Before Fukushima, there were about 440 operating nuclear reactors across the world, generating 14 percent of the world's electricity. The United States alone had 104 nuclear power plants. The risks from the spent fuel that those reactors generate is climbing year by year, as more reactors come online. China, South Korea, India, and other countries plan to build dozens more reactors. There is now a total global inventory of over one-quarter of a million tons of this fuel, with over 65,000 tons in the United States alone. Much of it is stored in poorly maintained and vulnerable pools.[6] "Will the world be safe with 1,000 nuclear reactors?" wondered former prime minister Naoto Kan after he left office. That question will reverberate in the years and decades ahead.

Kai Watanabe put in his service at the Daiichi plant and was moved early in 2012 after reaching the limit of his allowable radiation exposure. He clocked in for a few months at the Daini plant just down the coast. His parents live in Iwaki City, about 25 miles away, and have abandoned thoughts of returning home. The months Kai spent battling to bring the complex under control has scarred him in more ways than one. He is notably more cynical about the world and the people who run it. The once elite utility that dominated his life before March 11 has been shamed and bankrupted. But he retains his expansive, good-humored personality, joking that he is still waiting for his commemorative TEPCO pen.

Yoshio Ichida makes the short trip every day from his temporary home to the Pacific Ocean. The fishing cooperative remains a hollowed-out husk since the sea swept through its second-floor office and carried away everything inside. In the warehouse below the office, fishermen still gather for the daily ritual of testing their nets and equipment. Some take their boats out to sea to collect floating garbage and debris left over from the disaster, earning a stipend from the government. Once a month, Ichida and his colleagues meet to decide if they will fish again. The men listen to reports on radioactive readings taken from the sea, digesting once unfamiliar words like *cesium* and *becquerel*. In January 2012, groans greeted the news that nearly half of the 30 fish samples taken were found to be contaminated with radiation above government-set levels of 500 becquerels per kilogram. Plankton near the plant has been found with cesium up to 100 times predisaster levels. The fishermen suspect that rivers running into the sea from Fukushima are still contaminated with cesium. How long will it take to clear? At least Ichida still has his boat, but he worries that he may never be able to work again. In June 2012 some men were finally allowed to fish on a trial, limited basis.

Ichida's worries have been carried across the Pacific Ocean. Bluefin tuna caught off the California coast in August 2011 were contaminated with Fukushima's payload. Scientists found traces of two radioactive isotopes of the element cesium: cesium–137, with a half-life of about 30 years, was present in the eastern Pacific before the accident. But cesium–134 had not been detected in the Pacific for several years. This isotope has a half-life of about two years and is from man-made sources such as nuclear power plants and weapons. Scientists say that ingested or inhaled cesium can lodge in the body's soft tissue, increasing the risk of cancer. The scientists who reported the findings in May 2012 wrote that radioactive material found in the tuna was not considered dangerous for human consumption, as it was far less than the Japanese safety limit. But traces of cesium–134 were a surprise.[7]

The researchers concluded that the 15 tuna sampled must have been born near the plant at around the time of the accident. Bluefin tuna spawn only in the western Pacific off the coasts of Japan and the Philippines, and some migrate east to the California coast when they are young. From their size, the scientists knew that the young fish had left Japanese waters about a month after the accident. Daniel Madigan, lead author of the report, told Reuters that bluefin tuna near Japan soon after the accident could have had cesium–134 levels as much as 40 to 50 percent higher than normal.[8]

Concerns over food safety continue despite new tougher government limits for cesium in food and drink products that went into effect in April 2012. Under the new rules, the limit for general foods such as fruit, vegetables, rice, seafood, and meat is 100 becquerels of radiation per kilogram, down from 500 prior to April 1. The limit for milk, baby food, and infant formula is 50 becquerels per kilogram. For drinking water and tea leaves, it is 10 becquerels per kilogram.[9]

The new limits should help boost consumer confidence, and yet Japan has no centralized system to check for radiation contamination of food and drink products. Local municipalities and farmers are still responsible for carrying out testing.

Confusion and anger reigned during the months after the accident when unsafe levels of cesium were discovered in vegetables, seafood, and beef on supermarket shelves. In July 2011, the health ministry announced that hay contaminated with high levels of cesium had been fed to cattle before being shipped to meat markets. The meat was found to have as much as 2,300 becquerels of cesium per kilogram, nearly five times the government limit. Mushrooms, spinach, bamboo shoots, plums, fish, tea, and milk were found to be contaminated with cesium as far as 224 miles from the Fukushima plant.[10]

Mothers with small children were among the first to share information and pressure municipalities and food distributors to carefully monitor radiation testing. Their efforts sparked a grassroots movement. By December 2011, the newly formed National Network of Parents to Protect Children from Radiation included about 5,000 to 6,000 members from 250 individual groups. Among their activities was a petition drive handed to mayors and the education minister demanding that authorities ensure food safety in school lunches.

Ingredients in lunches served at public nursery, elementary, and junior high schools have become a major concern and the source of an uncomfortable debate within many families. Children have been caught in the middle when their parents or other family members had differing views on acceptable foods and cesium levels. Schools have allowed children to bring their own homemade lunches, but few have chosen to do so. Being different is not easy. Incidents of children being teased by classmates or even chastised by teachers have been reported in Fukushima.

In Sendai, a policy to reimburse lunch and milk fees to families of children who bring their own became official in April 2012. The city has about 80,000 children in public elementary and junior high schools where lunches and milk are served. About 580 have officially chosen the no-milk option. But only 61 students have chosen not to have the lunch or milk. This could be a reflection of more confidence in food testing or, perhaps, the fear of being ostracized. It is likely a combination of both.[11]

Concerned scientists such as physicist Dr. Ryugo Hayano have been promoting efforts to accurately measure radiation in school lunches as well as people's internal radiation contamination with whole body counters (WBC). From November 2011 to May 2012, he and his team measured about 10,000 people in the towns of Minamisoma and Hirata near the Fukushima Daiichi plant. The measurements showed that a large majority had no internal contamination. The levels were much lower compared with people in Russia, Ukraine, and Belarus five to ten years after the Chernobyl accident. He compares these findings with the higher average levels of internal contamination of Japanese people measured at the height of atomic bomb testing in 1964.[12]

Based on WBC measurements and further studies, Dr. Hayano expects that very few people in Fukushima will have more than 0.01 millisieverts of radiation per year. This is much lower than the 100 millisieverts of radiation per year risk level set by the government. Among his findings: Only 2 percent of the 53,000 food samples tested by local governments has exceeded the 100 becquerels of radiation per kilogram food regulation limit, no milk from Fukushima has yet been shown to be contaminated, and no contamination has been found in school lunches in Minamisoma since they started measuring them in January 2012. Dr. Hayano has concluded that food screening has been very effective, and "there is no health risk."[13]

These findings, however, have been questioned by experts including Dr. Ian Fairlie, a respected London-based independent consultant on radioactivity in the environment. "I hesitate to agree with Professor Ryugo Hayano's apparent conclusion that radiation levels in Fukushima residents are very low and no cause for alarm," he wrote in an e-mail. "He does not discuss the magnitude of the likely uncertainties of his internal dose estimates. These could be very large." Dr. Fairlie points to a 2004 report by the UK government's CERRIE

1 Sievert =	1,000 milliSievert = 1,000,000 microSievert	
6 - 10 Sievert	Single dose, fatal within weeks	
1 Sievert	Accumulated dosage estimated to cause fatal cancer in 5% of people (1)	*equivalent to 2.8 months living continuously at the Fukushima Daiichi Nuclear Power Plant*
2,579 microSievert per hour	Radiation level measured at the Fukushima Daiichi Nuclear Power Plant on March 20, 2011	
500 microSievert per hour	Mean radiation at the Fukushima Daiichi Nuclear Power Plant in 2011/04 (2)	
Approx. 0.05 microSievert per hour	Pre-accident mean radiation in Japan	
0.1 Sievert per year	Recommended limit for radiation workers averaged over five years (3)	
0.25 Sievert per year	Maximum annual dose allowable for workers at the Fukushima Daiichi Nuclear Power Plant during the crisis (4)	*equivalent to 21 days at the Fukushima Daiichi Nuclear Power Plant (*)*
2.7 microSievert per hour	Natural background radiation at airplane cruise altitude (5)	
27 microSievert	Tokyo - New York flight	*equivalent to 3 minutes at the Fukushima Daiichi Nuclear Power Plant (*)*
0.002 - 0.004 Sievert per hour	Average natural radiation per year (1)	*equivalent to 6 hours at the Fukushima Daiichi Nuclear Power Plant (*)*
10,000 - 30,000 microSievert	Full-body CT scan (5)	*equivalent to 40 hours at the Fukushima Daiichi Nuclear Power Plant (*)*
20 microSievert	Chest x-ray (6)	*equivalent to 2.5 minutes at the Fukushima Daiichi Nuclear Power Plant (*)*
5-10 microSievert	Dental x-ray (7)	*equivalent to 1 minute at the Fukushima Daiichi Nuclear Power Plant (*)*
(*)	Assuming 24 hrs exposure per day to the mean radiation at the Fukushima Daiichi Nuclear Power Plant in 2011/04	
	Mean radiation measurement at the Fukushima Daiichi Nuclear Power Plant was provided by TEPCO (2)	

Radiation Levels Explained

If the normal level of mean radiation in Japan is set at the width of a **credit card**, post-accident radiation at the Fukushima Daiichi Nuclear Power Plant (*) would be the height of the Empire State Building.

(Committee Examining Radiation Risks of Internal Emitters), a committee on internal radiation. The report warns about the many uncertainties that remain when estimating risks from internal radiation. He adds that Dr. Hayano's comparisons with human daily ingestion of potassium 40, a radioactive isotope, and the 1964 atomic bomb test exposures are inappropriate and "loaded" to give the impression that there is little to be concerned about.[14]

But Fairlie would surely agree with Hayano's findings that soil in Fukushima shows high levels of contamination. Food items such as wild berries that grow naturally in the soil and wild boar that eat the berries also show high levels. These, among other food sources affected by irradiated soil, will require monitoring for years to come, says Hayano.

As a former school cook and a grandmother, Setsuko Uwabe has been keenly aware of the issues. Her town, she said, is progressive about food safety even though it is far from the Fukushima plant. Nutritionists choosing ingredients for school meals avoid foods such as shiitake mushrooms, which are banned. At the city hall, devices that measure radioactive cesium in soil and food are available to the many residents who grow their own and who gather wild vegetables in nearby mountains. Irradiated soil and forest leaves remain a concern throughout the region.

Monitoring stations with do-it-yourself cesium-testing devices were a new concept in 2011 but may become the new normal. By December 2011, there were nine stations available in Fukushima and one in Tokyo. Each is run by a group that is part of the Citizens' Radioactivity Measuring Station consortium. Operating costs are covered by donations and fees charged for food testing—about ¥3,000 (US$38) per test. Test results are posted on each group's website.[15]

The timing of the accident was particularly bad for rice producers. The industry had been making inroads into the lucrative

Chinese market. Japanese rice had been gaining popularity among the wealthy, partly because it was considered safe. Japanese food imports are back to normal levels, and customers in the Tohoku region are gradually returning to locally grown produce. But Fukushima rice farmers face years of struggle. Their 2011 crop was deemed unsafe to consume. As we write, the 2012 crop has yet to be harvested, but hopes are high that it will be cesium-free. Farmers have sprinkled zeolite, a pebble-like material that traps cesium. They have added fertilizer with high levels of potassium to prevent seedlings from absorbing the cesium through their roots while growing. Worried farmers know that testing will be strict. Will the government buy and destroy the rice if it is contaminated again, as it did in 2011?[16]

The farmers know radiation levels will decline each year, but they realize that the challenges they face will be long term. Some farmers have given up on growing rice and turned to flower production, which does not require radiation checks. Others have taken on the mammoth task of suing TEPCO for damages.[17]

In 2012, 135 farms in Mayor Sakurai's Minamisoma were given special permission to plant rice on an experimental basis with the agreement that the harvest would be destroyed regardless of the cesium level. The farmers said they were doing everything they could to start growing rice again in 2013.[18] Beyond this, all they can do is hope and pray.

David Chumreonlert's trip to Kobe for a friend's wedding, just before he left briefly for the United States, turned out to be a life-changing event. There he met Mari Kikukawa, the Japanese woman he married in January 2012. The two have since moved to Sendai to be closer to his church, and David continues teaching English at the public schools there. But he misses Higashi-Matsushima—not only the nature and scenic views but also the friendliness and warmth of the local people and his many friends.

David feels strongly that he was saved from the tsunami for the purpose of helping others, which he found he could do through his church. The tiny congregation of about 23 members suddenly became a hub of international activity after the disaster. Affiliated church leaders and members, mainly from the United States and Taiwan, began streaming into Sendai to help with relief aid and rebuilding efforts.

As the onslaught of visitors continues, David remains the on-site translator, guide, coordinator, and cheerleader. The other church members, all Japanese, have been energized by the efforts and become more actively involved. The church received great interest during a recent gathering they organized at a temporary housing unit in Ishinomaki. David was surprised because most Japanese consider themselves Buddhist and do not generally follow Christian ideology. He sees it as a reaction to the daunting challenges they have faced since the disaster and the greater need for spiritual support and guidance.

Graduate school is a future goal. David would like to get a professional Japanese teaching license, and ideally, teach science. For now, his English teaching continues to be enjoyable, and the pay is solid. He is concerned about his job prospects if he moved back to the States, so he is glad to have the financial security in Japan.

David and Mari are discussing where to raise their family in the future. She would prefer Japan, but David feels that Texas would be a better environment. As an American with Thai roots and living in Japan with a Japanese wife, David will probably take any differences in stride, cultural or otherwise. After his near-death experience during the tsunami, his life now seems full of promise.

Setsuko Uwabe retired in September 2011 after working for 31 years as a cook for Rikuzentakata's public schools and nursery facilities. She does not miss working and enjoys getting together with her former colleagues on a regular basis. Her new life keeps her very busy.

In October that year, her son, Rigeru, decided to move back home from Sendai where he was working in advertising. He knew that his mother was having a hard time adjusting to life without Takuya. She was often alone, and he wanted to keep her company. But there was more. He wanted to bring his new partner and her two young boys, ages seven and nine, to live with her. Her house of mourning has become a hub of activity, laughter, and love. Setsuko admits with a smile that she is still adjusting to the lively pace of her new instant family.

During that first year after the disaster, Setsuko attended a series of memorial services for her husband and others in Rikuzentakata. For the first anniversary on March 11, 2012, Rigeru, along with his new wife and her two boys, visited the family grave at a Buddhist temple while Setsuko went to the town's large memorial service held at one of the public schools.

She felt oddly put out by this event, which was dominated by the presence of political bigwigs from the capital, among them Tokyo's governor, Shintaro Ishihara. The controversial Ishihara had called the quake/tsunami disaster "divine punishment" for Japan's modern "selfish greed." Remarkably, Tokyo's citizens reelected him for his fourth term a month later. The memorial event was supposed to help the city mourn for the dead, but it felt more like a public relations moment, with introductions that lasted way too long. But she was deeply moved by the emperor's speech, broadcast live on a large screen. His words matched her feelings exactly. Later there would be more controversy after some accused the television networks of censoring the speech when the emperor referred to the "formidable task" of overcoming the radioactive legacy of Fukushima.

When her life began settling down, she had more time to think of Takuya. It was not easy, as the memories came back stronger than ever. One day while driving through town, she happened to come upon the spot near the river where his body was found. That encounter, after so long avoiding the spot, took her time to get over.

In the spring of 2012, she decided to have a small cherry blossom viewing party and barbecue at home, like the many Takuya had organized with friends over the years. But after the guests had already been invited, she started to get cold feet. It suddenly seemed too soon, and she was afraid the gathering would depress her. She went ahead anyway and afterward was thankful she did. As she gazed into the bonfire that evening, she felt herself calmed and more at peace.

But she will forever carry the pain of losing Takuya so abruptly. Every morning after 6:00 A.M., she prays for him at the family's small home shrine, surrounded by his memorials. The house is still cluttered with his things. But soon she will move into a new cottage that will be built in a section of their large garden. She is planning an outdoor wooden deck and a space where she can invite friends and neighbors to socialize. Somehow, life is moving on.

Rikuzentakata is also rebuilding in fits and starts. Newly widowed Mayor Futoshi Toba is a single parent, struggling to raise his two sons while overseeing an eight-year reconstruction plan. He always brushes away praise and reacts uncomfortably to questions about grieving for his dead wife. Like Mayor Sakurai in Minamisoma, farther down the ruined Tohoku coast, his life is dedicated to his city and persuading its remaining 22,000 residents to stay. Mayor Toba keeps one wary eye on the drain of people from the city—about 1,000 have left so far. The next three years will be spent developing a design for the city and rebuilding infrastructure. Construction in the wrecked areas near the ocean has been banned. Mayor Toba hopes to turn the area into an ocean park for tourists.

New apartments are being built in an elevated location, and residents can apply for the units. Farmland and forests higher up will be converted into residential areas, but Setsuko worries that it will be inconvenient and difficult for the elderly to live there. A new temporary shopping mall opened in June 2012.

In 2012, the city's residents began planting cherry trees to mark the line where the tsunami waves reached. Hopefully they will remind future generations of the destructive power of the sea. The residents are trying to get permission from landowners to plant more. *Sakura* (cherry blossoms) are perhaps Japan's most potent symbol of the transience of life. For Setsuko and the many who lost loved ones in the tsunami, the symbolism strikes very close to home.

Toru Saito and his family were able to move out of the junior high evacuation site near their village in Oginohama on July 7, 2011, almost four months after the earthquake and tsunami. They moved to the city of Ishinomaki nearby and live in prefabricated temporary housing built for disaster victims. Their two-room unit is rent-free, and they will be able to stay for three years.

Although they have virtually no heating or cooling and no privacy, they agree it is much better than the evacuation center. There, they had to be extra cautious about not offending anyone and making sure relations were congenial. The experience was stressful for everyone.

Toru's two older brothers are busy working for the same Ishinomaki-based paper company and rarely have time to spend at the new unit. Toru is living near his Tohoku University campus in Sendai, about an hour away, and comes for short visits. The university waived tuition fees for students hit by the disaster and reimbursed the ¥280,000 (about US$3,500) admission fee. His mother is now a school bus attendant, and his father got a license to drive large vehicles and hopes to work as a truck driver.

In 2011, the family received about ¥2 million (about US$25,000) in three installments from a pool of private and public donations, as well as insurance on their house. But the insurance on their lumber factory did not cover natural disasters. There has been no further government money since then, and none is expected. Toru's mother says they are grateful for the temporary housing and feel this is enough. "We have to make the effort to stand on our own and not be dependent on government handouts."

Toru, however, thinks about the many who are unable to work and wonders how they can manage. "My parents are still healthy and can keep a job," he says. "But there are many who can't, and they need continued financial support." He cannot imagine that anyone is completely satisfied with the government's handling of the disaster and subsequent recovery efforts. So much more needs to be done to support those affected. It could take his entire lifetime.

Toru is enjoying his university life. He decided to major in mechanical engineering with a focus on robotics. He lives on his own in an apartment near campus and is trying his hand at cooking, with mixed results. A part-time job at a cram school teaching junior and senior high school students helps pay the rent.

Toru admits that he has no plans to move back to his beautiful village in Oginohama, though he misses it badly. Someday his children may have the chance to grow up in a similar place surrounded by nature. He is definite about one thing: he cannot live with the fear of earthquakes and tsunami.

Living outside Japan may be Toru's only choice. The country will always be wracked by earthquakes and pummeled by tsunamis. Since the March 2011 quake, concern is focused on Tokyo more than ever. In late March 2012, a government research team warned that Tokyo Bay, at the heart of the megalopolis, could be the epicenter for a quake that hits the top end of the Japanese seismic intensity scale. The following day a Cabinet Office panel announced that a magnitude 9 megaquake occurring in the Nankai Trough could rock areas stretching along the Pacific Coast, including Tokyo, and produce tsunami waves higher than 65 feet.[19]

Even if Toru moves abroad for work, refuge from quakes and tsunamis is not guaranteed, even in areas deemed safe. The unexpected 5.8 magnitude quake that shook the US East Coast in August 2011 not only rattled buildings and nerves, it triggered a serious look at quake-resistant building codes, safety measures, and preparedness. That it affected New York City and Washington, DC has made the rethink even more crucial. Cracks left in the Washington Monument and National Cathedral were a symbolic, eerie warning.

Other "surprises" in recent years have included the Indian Ocean megaquake and tsunami on December 26, 2004, that killed 230,000; the 2008 quake in China's Sichuan Province that took 68,000 lives; the 2010 Haiti earthquake with an estimated 316,000 deaths; and the 2010 and 2011 New Zealand quakes.

If Toru moved to idyllic Seattle or an area along the US Pacific northwest, he would join the millions living at risk. North America's Pacific Coast sits parallel to the Cascadia subduction zone. This is a 600-mile-long offshore fault that runs from northern California to southern British Columbia.[20]

Research has shown that large quakes along the Cascadia fault zone occur about every 250 to 500 years. The fault last shifted 312 years ago. Many scientists agree that a quake could produce the equivalent of the 2011 Japan tsunami that would reach the coast in about 20 minutes.[21]

Seattle might not be hit hard by a coastal megatsunami because it is located in Puget Sound behind the Olympic Peninsula. But a shallow fault has been discovered under the sound that could propel a tsunami right off of Seattle.[22] There is growing concern that the city and surrounding region are woefully unprepared.

Quake prediction would seem an obvious solution to saving lives and property. Yet it remains an illusive science, despite better understanding of plate tectonics. History has shown that factors determining the earth's geological movements are innately unpredictable. Since Japan's March 2011 megaquake, however, geologists and earthquake researchers worldwide have been able to reassess their assumptions with deeper understanding. One defining shift is the realization that megaquakes could occur on any subduction zone.

How do we minimize the risk of vast destruction, tragedy, and astronomical recovery costs? Simply put, by finding ways to predict when earthquakes occur. This would allow real tsunami prediction. Looking at geologic evidence of tsunami, such as paleo-tsunami deposits, will offer clues. Adhering to ancestral warnings and historical records are key. Combining short-term and long-term earthquake forecasts, new warning systems employing satellite and IT networks, all coordinated with regional and local authorities, would be an ideal answer. A follow-up answer: training entire populations on what to do when they hear the warning.

When and where will the next Big One happen? That answer may elude us but it is clear that there is much we can and should do to prepare.

NOTES

PROLOGUE

1. Sections of this account appeared in *The Irish Times* as "I'll tell him about the most terrifying two minutes of my life," March 12, 2011, http://www.irishtimes.com/newspaper/world/2011/0312/1224291982348.html.

CHAPTER 1: THE QUAKE

1. Quotations and other information on the six main characters in the book come from personal interviews conducted multiple times between March 2011 and July 2012, unless otherwise stated. The name Kai Watanabe is a pseudonym, used to protect his identity. Any other quotations not cited in the endnotes come from the authors' personal interviews. Translations of Kenji Miyazawa's poems are taken from the book *Strong in the Rain: Selected Poems*, trans. Roger Pulvers (Tarset in Northumberland, England: Bloodaxe Books, 2007).

2. Tokyo's controversial mayor, Shintaro Ishihara, famously described the quake and tsunami afterward as "divine punishment" for the "arrogance" of modern Japan. See Justin McCurry, "Tokyo Governor Apologises for Calling Tsunami 'Divine Punishment,'" *Guardian,* March 15, 2011, http://www.guardian.co.uk/world/2011/mar/15/tokyo-governor-tsunami-punishment.

3. Lucy Birmingham, "Japan's Earthquake Warning System Explained," *TIME,* March 18, 2011, http://www.time.com/time/world/article/0,8599,2059780,00.html. Magnitude measures the earthquake's shaking. The scale of the magnitude is logarithmic. If you add 1 to an earthquake's magnitude, you multiply the shaking by 10. For example, the 9.0 magnitude of the Great East Japan Earthquake in Tohoku was 10 times greater than the 8.0 magnitude earthquake in Sichuan, China, in 2008. For further explanation, see Robert Coontz, "Comparing Earthquakes, Explained," *ScienceInsider,* March 15, 2011, http://news.sciencemag.org/scienceinsider/2011/03/comparing-earthquakes-explained.html.

4. The speed of the tsunami was later estimated by *Scientific American*. See Francie Diep, "Fast Facts about the Japan Earthquake and Tsunami," *Scientific American,* March 14, 2011, http://www.scientificamerican.com/article.cfm?id=fast-facts-japan.

5. David McNeill, "Pro-Nuclear Professors Are Accused of Singing Industry's Tune in Japan," *Chronicle of Higher Education,* July 24, 2011, http://chronicle.com/article/Pro-Nuclear-Professors-Are/128382/. One of the earliest academic whistle-blowers, Hiroaki Koide, was held to assistant professor level for decades because of his opposition to nuclear power. See his *Genpatsu no uso* (*The Lies of Nuclear Power*) (Tokyo: Fusosha, 2011).

6. The American Nuclear Society, "Fukushima Daiichi: ANS Committee Report," March 2012, 28.

7. David McNeill, "Shaken to the Core: Japan's Nuclear Program Battered by Niigata Quake," *Japan Focus,* August 1, 2007, http://japanfocus.org/-David-McNeill/2487.

CHAPTER 2: TSUNAMI

1. Higashi-Matsushima is known for the Matsushima Airbase where the Japan Air Self-Defense Force trains. During the tsunami, the airbase was flooded with seawater, ruining the 18 F-2B fighter jets used for training.

2. Kyodo News, "To French, Japanese Towns, the World Is Their Oyster," *Japan Times,* May 2, 2012.

3. Bruce Parker, *The Power of the Sea* (New York: Palgrave Macmillan, 2010), 151–152.

4. Ibid.

5. Ibid. Used in Japan since the 1600s, the word *tsunami* literally means "harbor wave" and came from Japanese fishermen. When working offshore, they could not tell when a tsunami passed underneath their boat and would return to find their harbor destroyed. Detecting a tsunami in deep water is difficult because it is not high. It also has a slow up-and-down motion because of its long wavelength. Tsunamis are very different from storm-induced wind waves with high peaks, the kind that surfers ride, which are dangerous out at sea.

6. *The Chronicles of Japan* (*Nihon Shoki* or *Nihongi*), compiled in 720, is the second oldest official history of Japan. The oldest is the *Kojiki* (Record of Ancient Matters), dating from 711.

7. National Geographic Society, "Tsunamis: Killer Waves," *National Geographic,* accessed May 5, 2012, http://environment.nationalgeographic.com/environment/natural-disasters/tsunami-profile/.

8. Beth Rowen and Catherine McNiff, "Tsunami in Japan 2011: Waves Stirred Up by Earthquake Cause Wide Destruction," Infoplease, accessed May 5, 2012, http://www.infoplease.com/science/weather/japan-tsunami-2011.html#ixzz1tzQvGLmk.

9. Aislinn Laing, "Japan Earthquake: What Causes Them?" *Telegraph,* March 11, 2011, http://www.telegraph.co.uk/news/worldnews/asia/japan/8375788/Japan-earthquake-what-causes-them.html.

10. Rias are estuaries formed from a drowned river system caused by either the sea level having risen or the land having sunk. They usually occur along a rugged coast perpendicular to a mountain chain. They have a wide funnel shape and steadily increasing depth seaward, which can cause an exaggerated tidal effect and waves.

11. Masayuki Nakao, "The Great Meiji Sanriku Tsunami," *Failure Knowledge Database Project, Japan Science and Technology Agency,* March 2005, http://www.sozogaku.com/fkd/en/cfen/CA1000616.html; U.S. Geological

Survey (USGS), Historic Earthquakes, Sanriku Japan, March 2, 1933, 17:31 UTC, Magnitude 8.4, http://earthquake.usgs.gov/earthquakes/world/events/1933_03_02.php.

12. Kyodo News, "Tsunami Hit More than 100 Designated Evacuation Sites," *Japan Times,* April 14, 2011, http://www.japantimes.co.jp/text/nn20110414a4.html.

13. Japan Meteorological Agency, "Monitoring of Earthquakes, Tsunamis and Volcanic Activity," accessed March 20, 2012, http://www.jma.go.jp/jma/en/Activities/earthquake.html; *Yomiuri Shimbun,* "Revitalizing Japan: Building a Disaster-Resistant Nation / Tsunami Warning Systems to Be Built in Space, Sea Floor," *Daily Yomiuri,* February 3, 2012, http://www.yomiuri.co.jp/dy/national/T120202007018.htm. In response, Japan's government is now developing an "emergency tsunami warning system" that will enable direct tsunami observations and wave height forecasting.

14. Setsuko Kamiya, "Students Credit Survival to Disaster-Preparedness Drills," *Japan Times,* June 4, 2011.

15. Ibid.

16. Ibid.

17. Ibid.

18. There are three main ways of gathering information to predict the likelihood of tsunamis. One is researching tsunami catalogs of historical events. Another is determining the age of geologic deposits left by great earthquakes and tsunamis. And the third is creating computer simulations of tsunamis from potential great earthquakes and landslides around the world. U.S. Geological Survey, "Can It Happen Here?" USGS, October 31, 2011, http://earthquake.usgs.gov/learn/topics/canit.php.

19. Michael Welland, "Ignoring Tsunami Records: Hubris, Complacency, or Just Human Nature?" *Through the Sandglass* (blog), March 22, 2011, http://throughthesandglass.typepad.com/through_the_sandglass/2011/03/ignoring-tsunami-records-hubris-complacency-or-just-human-nature.html.

20. Reiji Yoshida, "869 Tohoku Tsunami Parallels Stun," *Japan Times,* March 11, 2012.

21. Daily Mail Reporter, "The Mystic Stone at Tsunami Tide's Highest Point That Saved Tiny Japanese Village from Deadly Wave," *Mail Online,* April 21, 2011, http://www.dailymail.co.uk/news/article-1379242/Japan-tsunami-Mystic-stone-tides-highest-point-saved-Aneyoshi-deadly-wave.html. The previous tsunami height record in Japan was 125.3 feet, which occurred also in Iwate Prefecture, near the city of Ofunato, during the 1896 Meiji Sanriku earthquake and tsunami.

22. NHK special broadcast, 14:25 UTC, March 12, 2011.

23. Zachary Cohen, "Scientists: Japan's Tsunami Broke Off Chunks of Antarctica," *NewsFeed, Time.com,* August 10, 2011, http://newsfeed.time.com/2011/08/10/scientists-japans-tsunami-broke-off-chunks-of-antarctica/.

CHAPTER 3: CLOSE THE GATE

1. Norimitsu Onishi, "Seawalls Offered Little Protection against Tsunami's Crushing Waves," *New York Times,* March 13, 2011, http://www.nytimes.com/2011/03/14/world/asia/14seawalls.html?pagewanted=all.

2. Takehiro Furuma, interview by Lucy Birmingham, February 15, 2012.

3. All yen/dollar conversions were done in May and June 2012.
4. Wamura Kotoku, *Bimbo to no Tatakai 40 Nen* (*A 40-Year Fight against Poverty*) (Tokyo: Kaisofuku, 2002).
5. Michishita Shigetada, interview by Lucy Birmingham, February 15, 2012.
6. Norimitsu Onishi, "Japan Revives a Sea Barrier That Failed to Hold," *New York Times,* November 2, 2011, http://www.nytimes.com/2011/11/03/world/asia/japan-revives-a-sea-barrier-that-failed-to-hold.html?pagewanted=all.
7. Ibid.
8. As of March 29, 2012. Confirmation with city of Kamaishi.
9. Norimitsu, "Japan Revives a Sea Barrier."
10. Danielle Demetriou, "Japan Tsunami Anniversary: Revisiting Rikuzentakata, the Town 'Wiped off the Map,'" *Telegraph,* March 10, 2012, http://www.telegraph.co.uk/news/worldnews/asia/japan/9134474/Japan-tsunami-anniversary-revisiting-Rikuzentakata-the-town-wiped-off-the-map.html.
11. As of April 5, 2012. Confirmation with town of Rikuzentakata.
12. Rob Gilhooly, "Island Fortresses Floated for Tohoku," *Japan Times,* March 6, 2012.
13. Ibid.; Naomi R. Pollock, "Kamaishi City Proposal: Toyo Ito & Associates, Architects," *Architectural Record,* March 2012, http://archrecord.construction.com/features/humanitarianDesign/Japan/Kamaishi-City-Proposal.asp.

CHAPTER 4: MELTDOWN

1. NHK special, "Merutodaun—Fukushima Daiichi Genpatsu anno toki nani ga" (*Meltdown—the Fukushima Daiichi Nuclear Plant: Behind the Scenes*), broadcast repeatedly in January 2012.
2. "Fukushima Daiichi: ANS Committee Report," The American Nuclear Society Special Committee on Fukushima, March 2012, p. 28, http://fukushima.ans.org/report/Fukushima_report.pdf.
3. David McNeill, "Sato Eisaku's Warning," *Japan Focus,* April 23, 2011, http://www.japanfocus.org/events/view/79. Sato wrote a book claiming that he had been toppled as governor, then framed on corruption charges, because of his growing opposition to nuclear power. See "Chiji masatsu tsurareta Fukushima-Ken oshoku jiken" (*Annihilating a Governor*) (Tokyo: Heibonsha, 2009). He and local people helped thwart TEPCO plans that experts say could have made the March 11 disaster much worse. A decade earlier, the company proposed to load hundreds of tons of mixed oxide fuel containing tons of plutonium. If it had succeeded, the fuel would have presented an even greater challenge "in terms of the threat of widespread and large-scale plutonium dispersal and devastating human health impacts," said Shaun Burnie, an independent nuclear analyst, in a personal email message, March 2012.
4. "Past 3,500 Years Saw Seven M9s," *Kyodo News,* January 27, 2012.
5. Kyodo News, "Tsunami Alert Softened Days before 3/11," *Japan Times,* February 27, 2012.
6. "Rethink of Tsunami Risk Was Way Too Late," Associated Press, February 22, 2012. A definitive warning had already been made in 2009. At a panel meeting on nuclear regulatory policy in June that year, tsunami

expert Yukinobu Okamura, with the government's National Institute of Advanced Industrial Science and Technology (AIST), insisted that new evidence of the massive Jogan earthquake and tsunami that hit the Tohoku coast in A.D. 869 be considered more closely. "I would like to ask why you have not touched on this at all," he demanded. "I find it unacceptable."

7. Hiroyuki Kawai, in interview by David McNeill, March 27, 2012. He added, "If [nuclear plants] have to be built in Japan, there must be meticulous care in running them. This wasn't the case here. TEPCO didn't raise its tsunami wall an inch. That's criminal negligence."

8. David McNeill and Nanako Otani, "Waiting for Doomsday: Living Next to 'the World's Most Dangerous Nuclear Power Plant,'" *Asia-Pacific Journal* 9, issue 19, no. 2 (May 9, 2011), http://www.japanfocus.org/site/view /3527.

9. The single generator that survived saved the idling reactors five and six from meltdown, it later emerged. The generator was nearly 43 feet above sea level.

10. David McNeill and Jake Adelstein, "Tepco's Darkest Secret," *Counter-Punch,* August 12–14, 2011, http://www.counterpunch.org/2011/08/12/tep cos-darkest-secret/.

11. "Panel: Wide Communication Gaps Hindered Response in Fukushima," *Asahi Shimbun,* December 27, 2011, http://ajw.asahi.com/article/behind _news/politics/AJ201112270046.

12. Mari Yamaguchi, "Nuke Evacuation Fatal for Old, Sick," Associated Press, March 10, 2012; "573 Deaths 'Related to Nuclear Crisis,'" *Yomiuri,* February 5, 2012, http://www.yomiuri.co.jp/dy/national/T120204003191.htm.

13. Norimitsu Onishi and Martin Fackler, "Japan Held Nuclear Data, Leaving Evacuees in Peril, *New York Times,* August 8, 2011, http://www.nytimes .com/2011/08/09/world/asia/09japan.html. The SPEEDI system can now be found online at http://www.bousai.ne.jp/eng/.

14. Mark Willacy, "Japan 'Betrayed Citizens' over Radiation Danger," ABC News, January 20, 2012, http://www.abc.net.au/news/2012-01-19/japan -delayed-radiation-details/3782110.

15. Naoto Kan, in interview by David McNeill, April 6, 2012. Also see "Inside Japan's Nuclear Meltdown," Frontline PBS, February 28, 2012, http:// www.pbs.org/wgbh/pages/frontline/health-science-technology/japans -nuclear-meltdown/naoto-kan-japan-was-invaded-by-an-invisible-enemy/.

16. Quoted in Takashi Hirose, *Fukushima Meltdown: The World's First Earthquake-Tsunami-Nuclear Disaster* (2011), Kindle edition. "Most of the media believed this and the university professors encouraged optimism. It makes no logical sense to say, as Edano did, that the safety of the containment vessel could be determined by monitoring the radiation dose rate. All he did was repeat the lecture given him by TEPCO." As media critic Takeda Tōru later wrote, the overwhelming strategy throughout the crisis, by both the authorities and big media, seemed to be reassuring people, not alerting them to possible dangers.

17. Japan's government later officially raised Fukushima to INES (International Nuclear and Radiological Event Scale) level 7—the same as the 1986 Ukraine disaster.

18. BBC, *Inside the Meltdown,* directed by Dan Edge, aired February 23, 2012. David McNeill was a consultant on this documentary.

CHAPTER 5: THE EMPEROR SPEAKS

1. David McNeill, "The Night Hell Fell from the Sky," *Japan Focus,* March 10, 2005, http://www.japanfocus.org/site/view/1581.
2. Richard Sisk and Helen Kennedy, "Japan on Verge of Nuclear Meltdown, but Heroic Workers Fight Reactor Fire to Stop Radiation Leaks," *New York Daily News,* March 17, 2011. Available online: http://articles.nydailynews.com/2011-03-17/news/29175711_1_fuel-rods-nuclear-plant-reactor/2 (accessed August 16, 2012).
3. The statement has since been taken down from the British Embassy website but it read: "We advise against all non-essential travel to Tokyo and north eastern Japan given the damage caused by the 11 March earthquake and resulting aftershocks and tsunami. Due to the evolving situation at the Fukushima nuclear facility and potential disruptions to the supply of goods, transport, communications, power and other infrastructure, British nationals currently in Tokyo and to the north of Tokyo should consider leaving the area." David McNeill, email and telephone communication with British Embassy in Tokyo, August 21, 2012.
4. The claim is made in, among other places, a Japanese-only book by Kevin Maher, a former director of the Japan Desk at the US State Department in Japan. Maher, who was sacked in March 2011 on a separate issue, claims he opposed the plan. See "Ketsudan dekinai Nippon," (The Japan That Can't Decide) (Tokyo: Bunshun Shincho, 2011).
5. "Fukushima No. 2 Plant Was 'Near Meltdown,'" *Daily Yomiuri,* February 10, 2012, http://www.yomiuri.co.jp/dy/national/T120209007089 .htm; Rebuild Japan Initiative Foundation, Fukushima Genpatsu jikoo dokiritsu kensho iinkai, choosa, kensho hookokusho (Independent Investigation Commission on the Fukushima Daiichi Nuclear Accident), March 2012; and Martin Fackler, "Tokyo Weighed Evacuating Tokyo in Nuclear Crisis," *New York Times,* February 27, 2002, http://www.nytimes .com/2012/02/28/world/asia/japan-considered-tokyo-evacuation-during -the-nuclear-crisis-report-says.html?_r=1.
6. The story of what took place on the morning of March 15 has been recreated from a personal interview with Naoto Kan on April 6, 2012, and from other sources, including TEPCO company records, which were subsequently published. See "Kan Blasted Tepco, Said No Retreat from No. 1," *Japan Times,* March 16, 2012, http://www.japantimes.co.jp/text /nn20120316a3.html.
7. "Kan Blasted Tepco, Said No Retreat from No. 1," *Japan Times,* March 16, 2012, http://www.japantimes.co.jp/text/nn20120316a3.html; Naoto Kan, interview by David McNeill, April 6, 2012.
8. While the cult-like power of the emperor system ended after World War II, when the constitutional role of Emperor Hirohito was changed, some people still revere the monarchy, so the rumor that Emperor Akihito (Hirohito's son) had left the city was disturbing to them.
9. Shinichi Saoshiro, "Somber Japan emperor makes unprecedented address to nation," *Reuters,* March 16, 2011, http://www.reuters.com/article/2011 /03/16/us-japan-quake-emperor-idUSTRE72F23520110316 (accessed August 19, 2012).
10. See John M. Glionna, "Japan's 'Nuclear Gypsies' Face Radioactive Peril at Power Plants," *Los Angeles Times,* December 4, 2011, http://articles .latimes.com/2011/dec/04/world/la-fg-japan-nuclear-gypsies-20111204.

11. David McNeill, "Suicide Squads Paid Huge Sums amid Fresh Fears for Nuclear Site," *Independent,* March 30, 2011, http://www.independent.co .uk/news/world/asia/suicide-squads-paid-huge-sums-amid-fresh-fears-for -nuclear-site-2256741.html; Paul Jobin, "Dying for TEPCO? Fukushima's Nuclear Contract Workers," *Asia-Pacific Journal* 9, issue 18, no. 3 (May 2, 2011), http://japanfocus.org/-Paul-Jobin/3523; and Jake Adelstein and Stephanie Nakajima, "Tepco: Will Someone Turn off the Lights?," *Atlantic Wire,* June 28, 2011, http://www.theatlanticwire.com/global /2011/06/tepco-will-someone-turn-lights/39364/.

12. Akihisa Shiozaki, interview by Craig Dale, Canadian Broadcasting Corporation, March 2012. See also interview with lead author of report, Yoichi Funabashi in *The Asahi Shimbun,* "Fukushima Nuclear Crisis Revealed Japan's Governing Defects," February 29, 2012. "While he excessively micromanaged, he also understood what the government had to do at the most vital time of the crisis and what decision had to be made at that time. At that time, Kan was correct." Available online at http://ajw.asahi.com /article/0311disaster/fukushima/AJ201202290078 (accessed August 20, 2012).

13. The National Diet of Japan, "The Official Report of the Fukushima Nuclear Accident Independent Investigation Commission," executive summary, 2012, p. 33.

14. David McNeill, "Report Urging Mass Evacuation of Tokyo Residents Kept Secret," *Irish Times,* January 27, 2012, http://www.irishtimes.com/news paper/world/2012/0127/1224310807601.html.

15. "Edano Voices Distrust of Tepco, Nisa," *Daily Yomiuri,* May 29, 2012, http://www.yomiuri.co.jp/dy/national/T120528003609.htm.

16. Jun Hongo, "Depth of Fukushima No.1 Evacuation plan unclear in videos," *Japan Times,* August 8, 2012, http://www.japantimes.co.jp/text /nn20120808a2.html (accessed August 20, 2012).

17. Ibid.

CHAPTER 6: TELLING THE WORLD

1. The original *Sun* article may be seen here: http://www.thesun.co.uk/sol /homepage/news/3473142/My-nightmare-trapped-in-post-tsunami-Tokyo -City-of-Ghosts.html. For the "Journalists Wall of Shame," see http://www .jpquake.info/home. See also, David McNeill, "Sensationalist Coverage," in *The Irish Times,* March 19, 2011, http://www.irishtimes.com/newspaper /world/2011/0319/1224292611835.html.

2. "Japan Criticizes Foreign Media's Fukushima Coverage," *Asahi,* April 9, 2011. The *Blade* has a daily circulation of about 168,000.

3. Takashi Yokota and Toshihiro Yamada, "Sono toki kisha wa nigeta" ("At That Time, the Journalists Ran Away"), *Newsweek Japan,* April 5, 2011. Published in English as "Foreign Media Create Secondary Disaster," *No. 1 Shimbun,* June 2011, http://www.fccj.ne.jp/no1/issue/pdf/June_2011.pdf.

4. Mariko Sanchanta, "Japan, Foreign Media Divide," *Wall Street Journal,* March 19, 2011, http://online.wsj.com/article/SB10001424052748703512 404576209043550725356.html?mod=WSJAsia.

5. The original article may be seen here: http://www.thesun.co.uk/sol/home page/news/3473142/My-nightmare-trapped-in-post-tsunami-Tokyo-City -of-Ghosts.html. For the "Journalists Wall of Shame," see http://www .jpquake.info/home. See also David McNeill, "Sensationalist Coverage,"

Irish Times, March 19, 2011, http://www.irishtimes.com/newspaper
/world/2011/0319/1224292611835.html.

6.   Jeff Kingston, interview by David McNeill, March 21, 2011.

7.   "'Hōshanō ga kuru' no hyōshi ni hihan, Aera ga shazai" ("Aera Apologizes
     After criticism of "Radiation is Coming" Cover"), *Yomiuri,* March 21,
     2011,   http://www.yomiuri.co.jp/national/news/20110320-OYT1T00786
     .htm.

8.   *Shūkan Shinchō* magazine dubbed the management of TEPCO *senpan* (war
     criminals). *Shūkan Gendai* magazine named and shamed the most culpable
     of Japan's elite pronuclear scientists, calling them *goyō gakusha* (govern-
     ment lackeys) and *tonchinkan*—roughly meaning "blundering idiots."

9.   Laurie A. Freeman, "Japan's Press Clubs as Information Cartels" (Work-
     ing Paper No. 18, Japan Policy Research Institute, University of San
     Francisco Center for the Pacific Rim, April 1996), http://www.jpri.org
     /publications/workingpapers/wp18.html. Also see Freeman's *Closing the
     Shop: Information Cartels and Japan's Mass Media* (Princeton, NJ: Prince-
     ton University Press, 2000); William DeLange, *A History of Japanese Jour-
     nalism: Japan's Press Club as the Last Obstacle to a Mature Press* (Japan:
     Japan Library, 1998).

10. "Time Out Meets the Journalist TEPCO Loves to Hate," April 3, 2011.
     Reposted on Uesugi's own website at http://uesugitakashi.com/?p=677.

11. Interview with M. Wakiyama, "The Media Is a Mouthpiece for Tepco,"
     No. 1 Shimbun, June 2011, http://www.fccj.ne.jp/no1/issue/pdf/June_2011
     .pdf.

12. http://www.nikkei-koken.com/, October 3, 2011; interview by David Mc-
     Neill with the Nikkei Advertising Research Institute, July 25, 2012.

13. For details, see T. Koizumi, "Genpatsu suishin PR sakusen no ichidoku-
     santan" ("A Reading of Pro-nuclear Power PR Strategies"), in *Daijishin
     genpatsu jiko to media* (The media and the earthquake / nuclear disaster),
     ed. Media Kenkyūjo (Tokyo: Otsuki Shoten, 2011).

14. According to *Shūkan Gendai* magazine, the utility spent roughly $26 mil-
     lion on advertising with the *Asahi* newspaper. Its quarterly magazine, *Sola,*
     was edited by former *Asahi* writers. "Skūpu repōto: saidai no tabū, Tōden
     mane to *Asahi* shinbun" ("The Biggest Taboo: Tepco's Money and the
     Asahi Newspaper"), *Shūkan Gendai,* August 22, 2011.

15. Takashi Hirose, *Fukushima Meltdown: The World's First Earthquake-
     Tsunami-Nuclear Disaster* (2011), Kindle edition, http://www.amazon
     .com/Fukushima-Meltdown-Earthquake-Tsunami-Nuclear-Disaster-ebook
     /dp/B005OD75J2. This is not to suggest that the media completely ignored
     nuclear power—just that the odds were heavily tilted against a balanced
     discussion.

16. By that time, a steady stream of foreign and freelance reporters had been
     to see the town (Agence France-Presse was the first to arrive on March
     18).

17. David McNeill, "Pro-Nuclear Professors Accused of Singing Industry's
     Tune in Japan," *Chronicle of Higher Education,* July 24, 2011.

18. Ellis Krauss (a professor of Japanese politics and policy making at the
     University of California), quoted in the *Washington Post* in Chico Har-
     lan, "In Japan, Disaster Coverage Is Measured, Not Breathless," March
     28, 2011, http://www.washingtonpost.com/lifestyle/style/in-japan-disaster
     -coverage-is-measured-not-breathless/2011/03/26/AFMmfxlB_story.html.

19. "Kensho: higashi Nippon dai shinsai to media," *Galac* (3.11 and the Social Media Revolution: An Investigative Report on the Tohoku Disaster), October 2011.

20. Ibid.

21. M. Fackler, "Japanese City's Cry Resonates around the World," *New York Times,* April 6, 2011; David McNeill, "A City Left to Fight for Survival after the Fukushima Nuclear Disaster," *Irish Times,* April 9, 2011.

22. Satoru Masuyama, interview by David McNeill, November 24, 2011; "Barriers to Coverage: High Hurdles Blocked Reporting of Fukushima Nuclear Accident," *Asahi Shimbun,* July 13, 2011, http://ajw.asahi.com /article/0311disaster/analysis/AJ201107134358.

23. "Inside the Danger Zone," *Daily Beast,* April 3, 2011, http://www.the dailybeast.com/newsweek/2011/04/03/inside-the-danger-zone.html; "Fear and Devastation on the Road to Japan's Nuclear Disaster Zone," *Independent,* March 26, 2011, http://www.independent.co.uk/news/world /asia/fear-and-devastation-on-the-road-to-japans-nuclear-disaster-zone -2253509.html; "In the Shadow of Japan's Wounded Nuclear Beast," *Irish Times,* March 28, 2011, http://www.irishtimes.com/newspaper/world /2011/0328/1224293221947.html; "Barriers to Coverage," *Asahi Shimbun*; Keiichi Satō, interview by Asahi Shimbun, November 28, 2011.

24. Jochen Legewie, *Japan's Media: Inside and Outside Powerbrokers* (Tokyo: Communications and Network Consulting Japan, 2010), http://www.cnc -communications.com/fileadmin/user_upload/Publications/2010_03 _Japans_Media_Booklet_2nd_Ed_JL.pdf]; Eriko Arita, "Rebel Spirit Writ Large," *Japan Times,* October 2, 2011, http://www.japantimes.co.jp/text /fl20111002x1.html. While writing this book, a scandal involving hidden losses at the camera and optical equipment maker Olympus was in full flow. The scandal was broken by a tiny subscription-only magazine called *FACTA,* whose editor, Abe Shigeo, quit his job at the Nikkei after being told to spike a story on corruption in the securities industry. "There is no investigative reporting at Japanese newspapers," he said. Personal interview with David McNeill, December 2011.

25. Freeman, "Japan's Press Clubs as Information Cartels"; David McNeill, "Japanese Journalism Is Collapsing," Foreign Correspondents' Club of Japan, March 17, 2010, http://www.fccj.or.jp/node/5491.

26. Teddy Jimbo, interview by David McNeill, September 16, 2011. Most of the big newspapers and networks in Japan also agreed early on to avoid using the word *meltdown* (全炉心溶融) and settle for *partially melting* (部分的溶融), although the decision was made after a lot of debate. It will also be noticed that very little appeared in the Japanese media about the plutonium fuel in reactor three of the Daiichi plant.

27. Richard Lloyd Parry, interview by David McNeill, October 6, 2011.

28. "Inside Report from Fukushima Nuclear Reactor Evacuation Zone," You Tube video, 12:05, posted by "videonewscom," April 6, 2011, http://www .youtube.com/watch?v=yp9iJ3pPuL8.

29. Teddy Jimbo, interview by David McNeill, December 2011.

CHAPTER 7: FLYJIN

1. The CBS *60 Minutes* episode "Disaster in Japan" was aired on March 20, 2011, http://www.cbsnews.com/video/watch/?id=7360240n.

2. Confirmed with US Embassy Tokyo press office, August 24, 2012.
3. Ibid. The US Embassy Tokyo press office did not deny that passengers were charged a fee for the services.
4. John Roos, US ambassador to Japan, interview by Lucy Birmingham, February 22, 2012.
5. Steven Mufson, "NRC Fukushima Transcripts Show Urgency, Confusion Early On," *Washington Post,* February 22, 2012, http://www.washington post.com/business/economy/nrc-fukushima-transcripts-show-urgency -confusion-early-on/2012/02/21/gIQAkPTFSR_story.html; Brian Wingfield and Jim Efstathiou, Jr., "Wider U.S. Evacuation at Fukushima Supported in NCR Transcripts," *Bloomberg News,* February 24, 2012, http://www .businessweek.com/news/2012-02-24/wider-u-s-evacuation-at-fukushima -supported-in-nrc-transcripts.html; U.S. Nuclear Regulators Commission, "Actions in Response to the Japan Nuclear Accident: Public Meetings/Presentations," 2012, http://www.nrc.gov/reactors/operating/ops-experience /japan/japan-meeting-briefing.html; SimplyInfo, "NRC Fukushima Transcripts," February 22, 2012, http://www.simplyinfo.org/?page_id=5016.
6. The transcripts were released by the NRC on February 22, 2012, in response to requests under the Freedom of Information Act from the Associated Press and other news organizations. The 3,000 pages include conversations from March 11 through March 20. Many have been redacted, including over 16 pages of dialogues held on March 12. US nuclear power experts have been debating the lessons learned from the Fukushima accident. The NRC subsequently approved a number of safety steps at US plants, specifically focusing on the 104 US commercial nuclear reactors similar to those at Fukushima.
7. David Jolly and Denise Grady, "Anxiety Up as Tokyo Issues Warning on Its Tap Water," *New York Times,* March 23, 2011, http://www.nytimes .com/2011/03/24/world/asia/24japan.html?pagewanted=all.
8. Kyodo News, "Global Migration Body Helps Quake-Stranded Foreigners Exit Japan," *Japan Times,* March 27, 2011.

CHAPTER 8: HELP US, PLEASE!

1. Alex Martin, "Military Flexes Relief Might, Gains Newfound Esteem," *Japan Times,* April 15, 2011, http://www.japantimes.co.jp/text /nn20110415f1.html#.UAWpT2B30Xw.
2. Kosuke Takahashi, "Scandals Strain US-Japan Relations," *Asia Times,* March 12, 2011, http://www.atimes.com/atimes/Japan/MC12Dh01.html.
3. Ian Sample, "Japan Earthquake and Tsunami: What Happened and Why," *Guardian,* March 11, 2011, http://www.guardian.co.uk/world/2011 /mar/11/japan-earthquake-tsunami-questions-answers; Francie Diep, "Fast Facts about the Japan Earthquake and Tsunami," *Scientific American,* March 14, 2011, http://www.scientificamerican.com/article.cfm?id=fast -facts-japan.
4. The 31st MEU is a US Marine Corps rapid-response force stationed in Okinawa Prefecture. Lieutenant Karl Hendler and Corporal Kevin Miller, interview by Lucy Birmingham, May 2011.
5. Lieutenant Karl Hendler, Corporal Kevin Miller, and Captain Caleb Eames, interview by Lucy Birmingham, May 2011.
6. Reiko (mother) and Wataru (son) Kikuta, interview by Lucy Birmingham, February 13, 2012.

7. Eric Johnston, "Operation Tomodachi a Huge Success, but Was It a One-Off?" *Japan Times,* March 3, 2012, http://www.japantimes.co.jp/text/nn 20120303f1.html; John Roos, US ambassador to Japan, interview by Lucy Birmingham, February 22, 2012.

8. Johnston, "Operation Tomodachi a Huge Success." There were 38 Americans at the Daiichi nuclear power plant when the earthquake hit, including nuclear technician Carl Pillitteri. He recounted his experience on a PBS news program a year after the accident: "Fukushima Survivor: 'I've Hardly Smiled This Whole Year,'" YouTube video, 6:38, posted by "PBSNews Hour," March 9, 2012, http://www.youtube.com/watch?v=qt4TvT83PJw &feature=relmfu.

9. Vice President Joe Biden, speech at Sendai Airport, Sendai, Japan, August 23, 2011, http://www.whitehouse.gov/the-press-office/2011/08/23 /remarks-vice-president-biden-sendai-airport; "Is U.S. Military Relief Effort Operation Tomodachi Really about Friendship? (Japan Today)," *Finance GreenWatch,* April 24, 2011, http://financegreenwatch.org/?p=844.

10. Chalmers Johnson, "Tomgram: Chalmers Johnson on Imperial Rights," TomDispatch.com, December 5, 2003, http://www.tomdispatch.com/post /1112/; David McNeill, "Japan and US Agree Okinawa Troop Withdrawal," *Irish Times,* April 28, 2012, http://www.irishtimes.com/newspaper /world/2012/0428/1224315293728.html.

11. Masami Ito, "Global Rescue Teams Arrive to Lend Hand," *Japan Times,* March 14, 2011, http://www.japantimes.co.jp/text/nn20110314a5.html# .UAWqZ2B30Xw; Masami Ito, "World Pitches in to Offer Support," *Japan Times,* March 31, 2011, http://www.japantimes.co.jp/text/nn20110331f2 .html#.UAWqxGB30Xw; "Disaster Donations Top 520 B. Yen; Volunteers Total 930,000," *Jiji Press,* March 6, 2012.

12. Ken Watanabe reading Kenji Miyazawa's poem "Strong in the Rain" with English subtitles of one translation, http://www.youtube.com/watch?v=i -EdVR53lVk&feature=related; Asian artists singing "Ame ni mo Makezu" ("Strong in the Rain"), Theme Song of Artistes 311 Love Beyond Borders, April 1, 2011, http://www.youtube.com/watch?v=1CCciATqpQ0; Sarah Berlow, "Cyndi Lauper: Trouper for Tohoku," *Japan Real Time* (blog of the *Wall Street Journal*), March 12, 2012, http://blogs.wsj.com /japanrealtime/2012/03/12/cyndi-lauper-trouper-for-tohoku/.

13. Sarajean Rossitto, interview by Lucy Birmingham, January 31, 2012. Sarajean explains there is little unity among organizations in Japan's nonprofit sector, but the five commonly cited categories include: (1) Japan-based international NGOs; (2) Japanese NPOs without overseas operations; (3) NPOs founded after the Kobe earthquake that dispatch volunteers to small-scale disasters; (4) volunteer groups that have sprung up since March 11; and (5) social entrepreneur groups, mostly Japanese NPOs, with a focus on socially responsible business or helping an organization start a business.

14. The process includes representatives from all the affected prefectures, several of Japan's ministries, and the Red Cross/Red Feather/Community Chest, which is also authorized to accept cash donations. There were also donations from overseas Red Cross affiliates. Many questions arose on how exactly to divide the money. Local assessments had to be made. A major challenge was the vast location covering 19 prefectures.

15. Randy Martin, interview by Lucy Birmingham, February 7, 2012.

16. Malka Older, interview by Lucy Birmingham, February 23, 2012.

CHAPTER 9: DEPARTURES

1. Futoshi Toba, mayor of Rikuzentakata, interview by Lucy Birmingham, May 2, 2011.
2. Setsuko's house is in one of Rikuzentakata's hilly neighborhoods and was not touched by the tsunami. But the massive earthquake and continuing aftershocks toppled furniture, and much was ruined.
3. Daisuke Wakabayashi and Toko Sekiguchi, "After Flood, Deaths Overpower Ritual," *Wall Street Journal,* March 22, 2011, http://online.wsj.com /article/SB10001424052748703858404576214361499201024.html.
4. Ibid.; Jiji Press, "Emperor, Empress Wish to Be Cremated in Simple Funerals," *Japan Times,* April 27, 2012, http://www.japantimes.co.jp/text /nn20120427a3.html#.UA5AQmB30Xw. For Christians in Japan, cremation is the norm, according to former mortician Shinmon Aoki.
5. Buddhism was "officially" introduced to Japan in the sixth century and became associated with funerals, rituals commemorating death, and cemeteries. Shinto is not a religion but Japan's ancient indigenous faith in which all deities of heaven and earth, such as mountains, trees, rocks, and islands, are worshipped. There is no absolute god, but multitudinous gods or spirits, and no founder or official scriptures. Shintoism is associated with birth, marriage, and celebratory occasions. Buddhist temples and Shinto shrines are often located near each other, their physical proximity a reflection of their complementary and intertwined practices.
6. Tomoko A. Hosaka, "Japan Tries to Find the Missing," Associated Press, April 26, 2011. Japan's Self-Defense Force units led most of the searches for months after the disasters, backed by police, coast guard, and US forces. Two undersea robots provided by the nonprofit International Rescue System Institute were also used. Search operations continued for more than one year.
7. Hiroko Nakata, "Japan's Funerals Deep-Rooted Mix of Ritual, Form," *Japan Times,* July 28, 2009, http://www.japantimes.co.jp/print/nn200907 28i1.html.
8. The family altar and grave are considered gateways to the next world through which spirits can return during the annual Bon festival. This festival is similar to the Christian All Souls' Day or Mexico's Day of the Dead. In Tohoku and northeastern Japan, Bon is celebrated on July 15, while in other parts of Japan, it is celebrated mainly on August 15.
9. The choice of a posthumous name (*kaimyō*) is not without controversy. Buddhist temples are known to charge prices reaching upward of ¥1 million (US$12,700) based on the level of donations.
10. Shinmon Aoki, interview by Lucy Birmingham, March 26, 2012. The 2008 film *Departures* (*Okuribito*) was directed by Yōjirō Takita. The 1984 black comedy film *The Funeral* (*Ososhiki*) directed by Juzo Itami offers a revealing look at the dark and light side of Japanese funerals.
11. The former empire of Japan occupied Manchuria from 1931 until the end of World War II.
12. Shinmon Aoki, *Coffinman: The Journal of a Buddhist Mortician* (Anaheim, CA: Buddhist Education Center, 2002), 78–79, 83. There is growing scientific evidence and debate that sightings and feelings during the near-death experience are controlled by a physiological reaction in the brain, a "neurological mechanism," rather than mystical or spiritual phenomena. Dean Mobbs and Caroline Watt, "There Is Nothing Paranormal about

Near-Death Experiences: How Neuroscience Can Explain Seeing Bright Lights, Meeting the Dead, or Being Convinced You Are One of Them," *Trends in Cognitive Science* 15, no. 10 (October 2011); Charles Q. Choi, "Peace of Mind: Near-Death Experiences Now Found to Have Scientific Explanations," *Scientific American*, September 12, 2011, http://www.scientific american.com/article.cfm?id=peace-of-mind-near-death.

13. Roger Pulvers, *Kenji Miyazawa: Strong in the Rain; Selected Poems* (Northumberland, UK: Bloodaxe Books, 2007), 46.

14. Rob Gilhooly, "Time Has Stopped for Parents of Dead and Missing Children," *Japan Times*, March 11, 2012, http://www.japantimes.co.jp/text /nn20120311f3.html.

15. Shingo Ito, "Monk Cares for Remains of Unknown Victims," *AFP-JIJI*, March 9, 2012. Japanese Buddhist tradition emphasizes the importance of retrieving the corpse of the deceased. It is believed that the spirit is trapped in this world if the body is unclaimed. More than 3,000 remained missing and about 500 recovered corpses were still unidentified as of the one-year anniversary.

16. Norio Akasaka, interview by Lucy Birmingham, May 9, 2012. *The Legends of Tono*, assembled in 1910 by Kunio Yanagita (1875–1962), is a folklore classic and must read for cultural insights on the Tohoku region. See also Inoue Hisashi, *New Tales of Tono*, trans. Christopher A. Robins (Portland, ME: Merwin Asia, 2012), http://merwinasia.com/books/forth coming/New_Tales_of_Tono.html. Some of Japan's popular spine-tingling horror films are rooted in Tohoku tales.

17. Hiroko Nakata, "Japan's Funerals Deep-Rooted Mix," *Japan Times*, July 28, 2009, http://www.japantimes.co.jp/print/nn20090728i1.html.

18. Reverend Paul Silverman, interview by Lucy Birmingham, March 24, 2012.

19. Richard Lloyd Parry, "Suicide Rates Are Increasing in Japanese Regions Most Effected [sic] by the Tsunami and Nuclear Disasters," *Australian*, June 17, 2011, http://www.theaustralian.com.au/in-depth/japan-tsunami /suicide-rates-in-japanese-region-most-effected-by-the-tsunami-and-nuclear -disasters-have-jumped/story-fn84naht-1226076940518; Rob Gilhooly, "Suicides Upping Casualties from Tohoku Catastrophe," *Japan Times*, June 23, 2011, http://www.japantimes.co.jp/text/nn20110623f1.html. There have also been suicides among SDF soldiers tasked with the gruesome job of retrieving the many corpses.

20. Father Alfons Deeken, interview by Lucy Birmingham, April 16, 2012; Eriko Arita, "Priest-Philosopher Makes Death His Life's Work," *Japan Times*, September 4, 2011, http://www.japantimes.co.jp/text/fl20110904x1.html.

21. Deeken interview. Brought up in a devout Christian family, Deeken's dying sister felt there was hope that she could be reunited again with her family in heaven. According to Christian tradition, the soul of the deceased goes to heaven and is united with God. There is no belief in reincarnation.

22. The association now has 40 chapters throughout Japan. Its three goals are to promote death education, to improve terminal care in hospitals and develop more hospice programs, and to establish mutual support groups for grieving people who have lost loved ones.

23. At that time, there was only one hospice in Japan. Now there are over two hundred.

24. Yukari Tachibana, interview by Lucy Birmingham, February 19, 2012.

25. Toshinao Kanno, interview by Lucy Birmingham, February 15, 2012.

26. Her name has been changed here for privacy reasons.

27. Brother Rodrigo Treviño, interview by Lucy Birmingham, April 7, 2012.

28. Lucy Birmingham, "Disaster Highlights Plight of Japanese Orphans," *TIME,* July 10, 2011, http://www.time.com/time/world/article/0,8599,208 1820,00.html. A system of government-backed support payments has been established for the orphans. Relatives taking care of the children are not foster parents but are considered legal guardians and do not have the full legal rights of a parent. Complicating the situation is the highly controversial Japanese system of child custody in divorce. Only one parent is granted legal custody.

29. Brother Treviño points out that 95 percent of children now living in Japanese foster care institutions have one or both parents living. About 50 percent are abuse cases, which increased dramatically after Japan signed the United Nations Convention of the Rights of the Child in 1994. There are 532 registered foster institutions in Japan. Among the 34 in the Tohoku region, nearly one-third are run by Catholic congregations. In Sendai there are five. La Salle Home is one of three that are Catholic based.

30. UNICEF defines "orphan" as a child who has lost one or both parents. http://www.unicef.org/media/media_45279.html.

31. Kiku Iwamoto, interview by Lucy Birmingham, March 5, 2012. "Comfort For Kids" was developed in partnership with Bright Horizons Family Solutions and JPMorgan Chase. The Dougy Center for Grieving Children and Families, another Portland, Oregon-based non-profit, was actively involved in its development.

32. Mao Sato, interview by Lucy Birmingham, March 1, 2012.

33. Social welfare offices in the region assigned nurses to visit temporary housing units to provide health checks, mainly for the elderly. But nurses have not been expected to offer psychosocial counseling. Counseling centers were first set up in the fall of 2011 near the housing units to help residents with questions and problems. Mao Sato notes that at the time, objectives were not yet concrete, and counselors were unsure of their roles. She stresses that the NGOs could have helped counsel the residents if they had been allowed to.

CHAPTER 10: TOHOKU DAMASHII

1. Translations of Kenji Miyazawa's poems are taken from the book *Strong in the Rain: Selected Poems,* trans. Roger Pulvers (Tarset in Northumberland, England: Bloodaxe Books, 2007).

2. The official National Police Agency Figure is 15,854 dead and 3,155 missing.

3. "Restoring Tohoku Fisheries," *Japan Times,* November 2, 2011, www.japan times.co.jp/text/ed20111102a2.html.

4. Personal interview, Shoichi Abe, head of the Soma fishing cooperative, August 19, 2011. Our interviews with six main characters were done repeatedly throughout 2011 and early 2012.

5. Ibid.

6. Energy News, "Japan Nuclear Expert: Humanity as a whole has literally never experienced something like Fukushima—'We will be fighting this radiation on the order of tens or hundreds of years,'" May 11, 2012, http:// enenews.com/japan-nuclear-expert-humanity-as-whole-has-literally-never -experienced-something-like-fukushima-we-will-be-fighting-this-radiation -on-the-order-of-tens-or-hundreds-of-years-video (accessed August 19, 2012).

7. American Nuclear Society Special Committee on Fukushima, "Fukushima Daiichi: ANS Committee Report," March 2012, p. 16, http://fukushima .ans.org/report/Fukushima_report.pdf.

8. Mizuho Aoki, "Tohoku Fears Nuke Crisis Evacuees Gone for Good," *Japan Times,* March 8, 2012, in *3.11 One Year On: A Special Report,* http://www.japantimes.co.jp/text/nn20120308f1.html. A cumulative dose of 100 millisieverts over a lifetime increases the risk of dying from cancer by 0.5 percent, according to the International Commission for Radiological Protection.

9. Kyodo News, "Fukushima to Test Milk from 10,000 Mothers," *Japan Times,* January 13, 2012, www.japantimes.co.jp/text/nn20120113a4.html.

10. David McNeill, "The Fight for Compensation: Lessons from the Disaster Zone," in Greenpeace, "Lessons from Fukushima," February 28, 2012, www .greenpeace.org/international/en/publications/Campaign-reports/Nuclear -reports/Lessons-from-Fukushima/.

11. Nassrine Azimi, "Cherry Blossoms in Fukushima," *International Herald Tribune,* May 16, 2012, http://www.nytimes.com/2012/05/16/opinion /cherry-blossoms-in-fukushima.html?_r=1.

12. David McNeill, "Japan Reveals Huge Size of Fukushima Cleanup," *Irish Times,* September 29, 2011, www.irishtimes.com/newspaper/world/2011 /0929/1224304933758.html.

13. Makiko Segawa, "After the Media Has Gone: Fukushima, Suicide and the Legacy of 3.11," *Asia-Pacific Journal* vol. 10, issue 19, no. 2 (May 7, 2012).

14. Norio Akasaka, interview by Lucy Birmingham, May 9, 2012.

15. Tokyo Denryoku Kabushiki Kaisha, Baishookin go, seikyusho, kojinsama-yoo (TEPCO compensation application form). Figures come from TEPCO, personal interview with Yoshikazu Nagai and Hiroki Kawamata, Corporate Communications Department, January 13, 2012.

16. "10% of Compensation Forms Filed/TEPCO's Arduous Application Process Blamed for Claimants' Slow Response," *Daily Yomiuri,* October 31, 2011, www.yomiuri.co.jp/dy/national/T111012005321.htm.

17. See Kazuaki Nagata, "Disaster Towns Left in Limbo: Mayors," *Japan Times,* January 16, 2012, www.japantimes.co.jp/text/nn20120116a3.html; David McNeill's interview with Professor Tim Mousseau in "Learning Lessons from Chernobyl to Fukushima," *CNNGO,* July 28, 2011, www .cnngo.com/tokyo/life/learning-lessons-chernobyl-fukushima-645874.

18. It also makes no provision for the many unexpected consequences of the disaster, such as the irradiation of a newly built apartment building in the prefecture, which used contaminated stones in its construction. Families inside the building will have to be relocated and the building likely destroyed. See Kyodo (news agency), "New Condo's Foundation Radioactive," *Japan Times,* January 17, 2012, http://www.japantimes.co.jp/text /nn20120117a1.html.

19. "The 38th Middle-Term Forecast," December 2, 2011, p. 3, http://www.jcer .or.jp/eng/economic/medium.html. Also see "TEPCO Seeks 690 Billion Yen More for Fukushima Compensation," *Asahi Shimbun,* December 27, 2011, http://ajw.asahi.com/article/0311disaster/fukushima/AJ201112270013; Tatsuyuki Kobori, "Fukushima Crisis Estimated to Cost from 5.7 Trillion Yen to 20 Trillion Yen," *Asahi Shimbun,* June 1, 2011, http://ajw.asahi .com/article/0311disaster/quake_tsunami/AJ201106010334; Japan Center for Economic Research, "Impact to Last Decade or More If Existing

Nuclear Plants Shut Down," April 25, 2011, p. 11, www.jcer.or.jp/eng/research/pdf/pe(iwata20110425)e.pdf.

20. Tomohiro Iwata, "TEPCO, Radioactive Substances Belong to Landowners, Not Us," *Asahi Shimbun,* November 24, 2011, http://ajw.asahi.com/article/behind_news/social_affairs/AJ201111240030.

21. H. T., "Tepco's Nationalization: State Power," *The Economist,* May 11, 2012, www.economist.com/node/21554735.

22. The 21st Century Public Policy Institute, a think tank affiliated with the Keidanren business lobby, warned in May 2012 that without major reform, Japan could fall out of the league of developed nations. "Unless something is done, we are afraid Japan will . . . again become a tiny country in the Far East," its report said. See Kevin Rafferty, "Inviting Economic Suicide," *Japan Times,* May 2, 2012, http://www.japantimes.co.jp/text/eo20120502a1.html.

23. Huw Griffith, "Japan PM: No Individual to Blame for Fukushima," AFP, May 3, 2012, http://www.google.com/hostednews/afp/article/ALeqM5ibUo1F9_HHBAR4-ZyT1Fv4PGzlKA?docId=CNG.ccbdca9c1d32e1a2e21cc3ea00808e2f.191.

24. Hiroko Tabuchi, "Rise in Oil Imports Drives a Rare Trade Deficit in Japan," *New York Times,* January 24, 2012, www.nytimes.com/2012/01/25/business/global/rise-in-oil-imports-drives-a-rare-trade-deficit-in-japan.html. Also see AP, "Green Energy's Future Murky as Summer Looms," *Japan Times,* May 10, 2012, http://www.japantimes.co.jp/text/nn20120510f2.html; Kyodo News, "Fuel Imports Put Trade Deficit Near ¥3 Trillion," *Japan Times,* July 26, 2012, http://www.japantimes.co.jp/text/nb20120726a1.html.

25. Norio Akasaka, interview by Lucy Birmingham, May 9, 2012.

26. Nobuyoshi Ito, interview by David McNeill, January 12, 2012. Also see Jiji, "Fukushima Resort to Tap Spa for Own Power, Revival Pitch," May 10, 2012, http://www.japantimes.co.jp/text/nn20120510f3.html.

27. Hideyuki Miura, "Healthy Forests Key to Rebuilding Tohoku Communities," April 3, 2012, http://ajw.asahi.com/article/0311disaster/opinion/AJ201204030019.

28. Miyazawa, "Strong in the Rain," translated in Pulvers, p. 24.

29. Miyazawa, "Stop Working," translated in Pulvers, p. 33.

EPILOGUE

1. A terabecquerel is a measurement of radiation, named after French physicist Antoine H. Becquerel (1852–1908). Kyodo News, "Fukushima Meltdowns' March 2011 Fallout Higher than Estimated, Near 900,000 Terabecquerels: Tepco," *Japan Times,* May 25, 2012, http://www.japantimes.co.jp/text/nn20120525b6.html.

2. For a very useful survey of the debate, see Matthew Penney and Mark Selden, "What Price the Fukushima Meltdown? Comparing Chernobyl and Fukushima," *Asia-Pacific Journal* 9, issue 21, no. 3 (May 23, 2011), http://japanfocus.org/-Mark-Selden/3535.

3. Shaun Burnie, Matsumura Akio, and Mitsuhei Murata, "The Highest Risk: Problems of Radiation at Reaction Unit 4, Fukushima Daiichi," *Asia-Pacific Journal* 10, issue 17, no. 4, http://japanfocus.org/-Murata-Mitsuhei/3742.

4. Kyodo News, "Oi Assembly Says Yes to Restarting Reactors," *Japan Times,* May 15, 2012, http://www.japantimes.co.jp/text/nn20120515a1.html.

5. Yuri Kageyama, "Nuclear arms advocates get bolder amid energy debate," *Japan Times,* August 3, 2012, http://www.japantimes.co.jp/text/nn20120803f1.html (accessed August 20, 2012).

6. Burnie, Matsumura, and Murata, "The Highest Risk."

7. Reuters, "Fukushima Radiation Seen in Tuna off California," *Asahi Shimbun,* May 29, 2012, http://ajw.asahi.com/article/0311disaster/fukushima/AJ201205290016.

8. Ibid.

9. Radiation in food is measured in becquerels, a gauge of the strength of radioactivity in materials such as iodine–131, cesium–134, and cesium–137. Exposure to radiation from these and other radionuclides involves the release of atomic energy that can damage human cells and DNA. According to the World Nuclear Association, prolonged exposure can cause leukemia and other forms of cancer. See http://www.world-nuclear.org/education/ral.htm. Children and babies are more susceptible.

10. Aya Takada, "Japan's Food-Chain Threat Multiplies as Fukushima Radiation Spreads," *Bloomberg,* July 25, 2011, http://www.bloomberg.com/news/2011-07-24/threat-to-japanese-food-chain-multiplies-as-cesium-contamination-spreads.html.

11. Makiko Kimura (whose son attends a junior high school in Sendai), interview by Lucy Birmingham, June 3, 2012.

12. Dr. Ryu Hayano's PowerPoint presentation, "Internal contamination of Fukushima citizens: What we have learned from the recent whole body counter measurements," given at Temple University in Tokyo on July 3, 2012, http://www.slideshare.net/safecast/temple-u-20120703.

13. For helpful information on radiation terminology, see William Post, "Facts and Information about Radiation Exposure," *True North Reports.com,* March 21, 2011, http://truenorthreports.com/facts-and-information-about-radiation-exposure; David L. Chandler, "Explained: Rad, Rem, Sieverts, Becquerels: A Guide to Terminology about Radiation Exposure," *MIT News,* March 28, 2011, http://web.mit.edu/newsoffice/2011/explained-radioactivity-0328.html.

14. The Committee Examining Radiation Risks of Internal Emitters (CERRIE) operated between October 2001 and October 2004, http://www.cerrie.org/about.php.

15. Mizuho Aoki, "Mothers First to Shed Food-Safety Complacency," *Japan Times,* January 4, 2012, http://www.japantimes.co.jp/text/nn20120104f1.html.

16. Yuri Kageyama, "Fukushima Farmers Pray for Cesium-Free Rice," *Japan Times,* June 1, 2012, www.japantimes.co.jp/text/nn20120601f3.html.

17. Ibid.

18. Ibid.

19. Kyodo News, "Nankai Quake Scenario Menaces Pacific Coast," *Japan Times,* April 1, 2012, http://www.japantimes.co.jp/text/nn20120401a1.html.

20. Tim Folger, "The Calm before the Wave," *National Geographic,* February 2012, 71.

21. Ibid.

22. Ibid.

INDEX